Max Julius Louis Le Blanc, Willis Rodney Whitney

The Elements of Electrochemistry

Max Julius Louis Le Blanc, Willis Rodney Whitney

The Elements of Electrochemistry

ISBN/EAN: 9783337276522

Printed in Europe, USA, Canada, Australia, Japan

Cover: Foto ©berggeist007 / pixelio.de

More available books at **www.hansebooks.com**

OF

ELECTROCHEMISTRY

BY

MAX LE BLANC

PROFESSOR OF CHEMISTRY IN THE UNIVERSITY OF LEIPZIG

TRANSLATED BY

W. R. WHITNEY

INSTRUCTOR OF CHEMISTRY IN THE MASSACHUSETTS INSTITUTE OF
TECHNOLOGY OF BOSTON, U.S.A.

London

MACMILLAN AND CO., Ltd.

NEW YORK: THE MACMILLAN CO.

1896

TRANSLATOR'S PREFACE

WITH the exception of a few alterations, either suggested or accepted by Professor Le Blanc, the present work is as nearly as practicable a literal translation.

The rapid advance of the subject within the past few years has rendered treatises which adequately present it scarce, therefore I hope to have satisfied a want, and to have increased the facility of study for the English reader in this field.

It is with pleasure that I express here my great indebtedness to Mr. J. A. Craw of Glasgow, Scotland, for the aid he kindly gave me in the translation, and to Professor A. A. Noyes of the Massachusetts Institute of Technology, Boston, U.S.A., for correcting the proofs.

W. R. WHITNEY.

AUTHOR'S PREFACE

The greater part of the present work was written during the winter term 1894-95 in connection with a course of lectures which I was then delivering. It is intended for students of science and those who, having studied the subject, are already in practice, as well as for those interested in electrochemistry. I have endeavoured to write as clearly and simply as possible, but for those whose previous knowledge of the subject is very slight, a careful study of the book is necessary before its maximum utility can be reached. There are certain methods of conception used in modern electrochemistry which the student must make his own, and this process does not take place without study.

The book presents a view of the present state of the subject, and may contain some new ideas. The references to the literature are limited to the most important articles. It would scarcely have been possible for me to write the book without personal

contact with Prof. Ostwald and access to his papers
on electrochemistry. The dedication of the book to
him is an expression of my gratitude.

Finally, I must not neglect to thank Dr. A. Dahms,
Dr. M. Trautscholdt, and Dr. J. Wagner for their kind
aid in reading the proofs.

M. LE BLANC.

LEIPZIG, *end of September* 1895.

CONTENTS

CHAPTER I

CHAPTER II

CHAPTER III

CHAPTER IV

CHAPTER V

CHAPTER VI

CHAPTER VII

CHAPTER VIII

CHAPTER I

Energy. — A clear conception of the fundamental
principles of energy is essential to a successful study
of electrochemistry, consequently we shall first consider the different forms of energy and their relations
to each other.

The important rôle which energy plays in human
affairs is well known. If we buy coal or an article
of food, the important point to be taken into consideration is, in reality, the quantity of energy we are
obtaining. The same holds true if the purchased substance is the electric current. The quantity of electrical
energy which the electric current gives us, determines
its cost.

We recognise five distinct kinds of energy, as
follows :—

1. Mechanical Energy.
2. Heat Energy.
3. Electrical Energy.
4. Chemical Energy.
5. Radiant Energy.

These different forms of energy are capable of
changing, one into another. For some of them arbitrary

B

units have long been accepted. In the case of mechanical energy, for instance, the unit commonly employed in technical applications is that quantity of energy which is expended in raising a gram-weight one centimeter high.

For the scientific measurement and expression of quantities of mechanical energy, the centimeter-gram-second system is in common use. According to this system the unit of work, the erg, is the work which is expended in moving the unit of mass (the mass of a gram), the unit distance (the centimeter), against the unit of force (the dyne).

The dyne or unit of force has been chosen as that force which, in one second, produces in the mass of one gram an acceleration of one centimeter. Gravitation acts upon the gram-mass, producing an acceleration of 980·6 centimeters, consequently the force of gravity acting upon that mass amounts to 980·6 dynes. The difference between the mass of a gram and the weight of a gram must be kept in mind. The former is invariable, and its unit is the mass of a cube of water one centimeter on each edge when at 4° C. The mass of any body, which, acted upon by the same force, received the same acceleration as the above mass of water, could serve as the unit of mass.[1]

The weight of any body, on the contrary, depends on its position, and, in general, becomes greater or less as the body is moved nearer to or farther from the earth, although the mass of the body does not change.

The gram-weight represents, then, that force with which the gram-mass is attracted towards the earth,

[1] As a matter of fact, a certain piece of platinum preserved in Paris, which is about a thousand times as great as the above-described unit, serves as unit of mass.

and since this is capable of imparting to a body a mean acceleration of 980·6 cm. per second, we say that the gram-weight is equal to 980·6 dynes, and the technical unit of work 1 gm. cm. = 980·6 dynes cm. = 980·6 ergs. Possessing such a system of units, we can express the quantities of mechanical energy in given cases, and can compare them with one another. The unit of quantity which has been chosen for heat energy is the hundredth part of the heat which is necessary in order to raise the temperature of one gram of water from zero to one hundred degrees centigrade.

After having chosen units for two kinds of energy, we are able, aided by the law of the conservation of energy, to determine how many units of the one kind are equivalent to one of the other. By experiment, it has been learned that 43280 gm. cm. = 42440 × 10³ ergs, changed into heat, produce the heat-unit mentioned above, which has received the name calorie. Consequently, this quantity is called the *mechanical equivalent of heat*. We might proceed in a similar manner with all the five kinds of energy, but practically, the electrical energy is the only other one for which units have as yet been established. It is possible then to determine, besides a mechanical equivalent of heat, an electrical equivalent of heat and a mechanical-electrical equivalent. We shall learn more about these values later.

We are satisfied at present to accept the fact of the changes of energy from one form to another without raising the question as to the circumstances under which these changes take place, or as to the conditions of equilibrium.

We will first study the case where two systems possessing different amounts of the same kind of energy are so arranged that the energy of one may pass into the other. Let us apply this consideration to the volume-energy of two gases; volume-energy being a kind of mechanical energy, we may measure it in the above-described mechanical units.

If we have a mass of gas in a closed vessel, we say that the gas possesses a certain amount of volume-energy, because in expanding it is capable of performing work.[1] Imagine a vessel having the form given in the cut standing in a vacuum and containing a movable piston, A, weighing 100 grams; if now, by the expansion of the gas, the piston be raised from a to b, the distance being 50 cm., then by means of the volume-energy of the gas 100 grams have been raised 50 cm., that is, 100 times 50 or 5000 units of work have been produced; consequently, the volume-energy of the gas has been decreased by this amount. If the piston had an area of 100 sq. cm., the unit of its area would weigh one gram, and we should say that the piston exerted a pressure p of one gram. The volume v by which the gas has been increased in this movement of the piston is 5000 cc., the product pv, expressed in grams and cubic centimeters, is also 5000, or the product pv gives us the number of units of work which were produced.

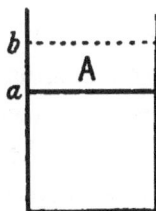

FIG. 1.

Imagine a horizontal vessel, as in Fig. 2, arranged

---

[1] It may remove a source of error to add here that the work which can be produced by the expansion of a gas is not derived from the internal energy of the gas. The gas is only the medium or agent which changes heat from the surroundings into work. If we say that a gas has a certain amount of volume-energy, we mean simply that the gas is capable of producing an equivalent amount of mechanical energy at the expense of the heat of its surroundings. Keeping this in mind, we may consider, for simplicity, that the volume-energy is possessed by the gas itself.

with a movable piston C, containing hydrogen on the left and nitrogen on the right of the piston. If the gases exerted equal pressure upon the piston it would remain at rest. There would be no passage of energy from one of the gases to the other.

FIG. 2.

The transference of energy is thus independent of the absolute quantities of energy which come into contact, since the gas filling the larger space has a greater quantity of volume-energy than the other. This difference in quantity may be made as great as desired by a proper choice of relative volumes; but if we change the density of one of the gases and consequently its pressure, the piston is set in motion, the volume of the denser gas increases, the gas loses volume-energy, while the volume of the other gas is diminished and its volume-energy increased. Equilibrium will again exist when the pressure exerted upon the piston by both gases has become the same.

Representing energy in general by E, the volume-energy of any body is expressed by the equation $E = pv$. To the factor p belongs, as we have seen, the important property of determining the equilibrium, and we call this the intensity-factor. The other quantity v is then simply equal to $\frac{\text{Energy}}{\text{Intensity}}$. It determines the amount of energy which, at a given intensity, exists in a system, and is called the capacity-factor. It is in this case evidently the volume.

It has been possible to decompose several of the forms of energy into two such factors—capacity- or quantity- and intensity-factors,—and this greatly aids in an understanding of energy phenomena.

**Electromotive Force, Current-Strength, Resist-
ance.** — Electrical energy is to be considered as the
product of the two factors : electromotive force (potential or tension), (π), and quantity of electricity (ϵ).
(Here the distinction between quantity of electricity
(ϵ) and the energy (E) is evident.)   The former quantity represents the intensity-factor, and the latter the
capacity-factor.   This will be made clearer in the
following pages.

On account of our limited sense of perception of
electrical phenomena, we are not in position to
comprehend them to the extent possible in the case
of mechanical energy.   The action and effects of
electrical energy must first be experimentally studied.
The imagination would not be able to grasp the idea
of the unit of work, or, let us say, of a meter, if the
action of the unit of work had not first been learned,
or if the length, which is represented by a meter,
had not been observed.

If we take a vessel which is divided into two parts
by a porous plate, as one made of unglazed porcelain,
and pour into one part a solution of copper sulphate,
and into the other a zinc sulphate solution, then put a
strip of copper into the copper and a strip of zinc into
the zinc solution, we have an arrangement called a
galvanic element.

If we connect the zinc and copper strips (the two
poles of the element) by means of a wire, the wire
becomes heated.   If we bring a magnetic needle near
it, the needle is turned from its natural position.
Finally, if we cut the wire, fasten a piece of platinum foil to each of the two ends, and dip these pieces
of foil into a copper sulphate solution in such a manner
that they are not in contact with each other, we observe

that metallic copper separates upon one of the pieces of platinum.

From these observations we must conclude that in this connecting wire some process takes place, for we have observed effects which were not observable before we united the zinc and copper with the wire.   When such effects are produced as here observed, we say that an electric current is passing through the wire.   It is conceivable that we might have a case in which the wire would affect the magnetic needle, but would not be heated, or would not possess all the properties peculiar to the electric current.   This was formerly supposed by many to be true, but, as a matter of fact, such is not the case.   We know from long experience that if a wire exhibits one of the above three phenomena, it also exhibits the other two, as well as a number of others which are not of interest here.   That many of the phenomena may be made to disappear under the conditions of the observations does not contradict the above statement.   We are now able by proper arrangements to ascertain the properties of the electric current.

If in a galvanic element, as previously arranged, we simply change the end connections so that the end which was formerly joined to the zinc is now joined to the copper, and the other end now joined to the zinc, we observe the same phenomena, with the simple difference that the magnetic needle is influenced so as to move in the opposite direction and that the copper is precipitated upon the other piece of platinum ; consequently we may properly speak of the direction of the electric current.

Naturally, the next thing to observe is whether the deflection of the magnetic needle or the amount of copper separating out in a given time always remains the same, and upon what the variation depends, if there is any. To this end we lengthen the connecting wire, and observe that the rate of the precipitation of copper is decreased, while by shortening the wire the rate becomes greater. We therefore conclude that the electric current has a strength dependent on circumstances, and we obtain an idea of *current-strength*. The current-strength has been decreased by increasing the length of the wire, and increased by shortening; therefore the wire hinders, to a certain extent, the passage of the current, *i.e.* the wire possesses a certain *resistance*. We have found that the greater the resistance, the less is the current-strength. Now the question arises : Is it possible to change the current - strength without altering the resistance? Experiment answers, Yes. If, instead of using one electrical element, we use two, the zinc of one connected with the copper of the other, we obtain a much greater current-strength, although the resistance of the circuit has been increased by the introduction of the second element. Here the effect is as if the pressure under which the electric current is driven through the wire had been increased, and consequently we come to speak of the *electromotive force.*

We may now assume that the words current-strength, resistance, and electromotive force are not meaningless terms to the reader, but that their use is understood. We must now proceed to study the units of these quantities, and in doing so we shall follow a simpler way than that by which these units were established. The electromotive force of the previously described element (named, from its discoverer, the Daniell element), the concentrations of the two solutions being alike, we place at 1·10 unit, and give to the unit the name *volt*. For the unit of resistance we use that resistance possessed by a column of mercury 106·3 cm. long, and of one sq. mm. cross-section at zero degrees. This unit is called an *ohm*. By the unit of current-strength we mean that current by which 0·328 mg. of copper are precipitated in a second. This is called an *ampère*.[1] Why just these quantities have been accepted as units need not occupy our attention here; it is a question belonging more to the history of the subject.

We already know that the current-strength is dependent upon the electromotive force on the one hand, and upon the resistance on the other. Ohm made the assumption that the current-strength is directly proportional to the electromotive force, and inversely proportional to the resistance. This assumption has been proved correct. We may write

$$\text{current-strength} = \frac{\text{electromotive force}}{\text{resistance}}\, k,$$

where k is a ratio-factor dependent upon the chosen units, but we have here chosen the units, so that if there exists in a circuit whose resistance is one ohm an electromotive force of one volt, the current-strength is exactly one ampère. Accordingly

$$\text{ampère} = \frac{\text{volt}}{\text{ohm}},$$

the factor k in this case being 1. If we had chosen

---

[1] These terms, as well as the coulomb and farad (explained later), have been derived from the names Volta, Ohm, Ampère, Coulomb, and Faraday, men whom we may call the pioneers of the science of electricity.

a unit ten times as great for current-strength, k would have been 0·1.

We are now in a position to see how unknown electromotive forces and resistances are determined. It is evident that to determine the current-strength it is only necessary to ascertain the number of milligrams of copper precipitated in a second, and divide this number by 0·328 ; the quotient is the current-strength in ampères. If we wish to determine the resistance of the circuit, we may take a Daniell element possessing an electromotive force of 1·10 volt, and measure the current-strength which it produces in the circuit. Let us say that we obtain 0·001 ampère, then according to Ohm's law the resistance must be 1100 ohms.

$$\frac{1\cdot10}{0\cdot001} = 1100 \text{ ohms}$$

$\left(C = \dfrac{\pi}{R} ; \quad C = \text{current - strength} ; \quad \pi = \text{electromotive} \right.$

force ; $R = $ resistance ; consequently $\left. R = \dfrac{\pi}{C} \right).$

If now we connect into the same circuit, instead of the Daniell, an unknown electromotive force (π), and do not alter the resistance, we can easily learn the value of π in volts by measuring the new current-strength. Let us say, for example, that we have here a current-strength of $\frac{1}{100}$ ampère, then the electromotive force is $\pi = \frac{1}{100} \cdot 1100 = 11\cdot0$ volts.

In order to obtain still clearer ideas of the electric current, let us consider its analogy to a stream of water. Electromotive force corresponds to the pressure of the water, the electrical resistance to the friction-resistance, and the strength of the electric current to the current-strength or rate of flow of the stream of water. When we say that a stream possesses a certain current, we mean that in a unit of time a certain quantity of water passes through a cross-section. A unit for current of water has not been established for scientific use. We might consider as a unit the current by which one cubic meter passes in a second.

Just as we speak of the quantity of water in the stream, we may also speak of the quantity of electricity in the electric current, without necessarily imagining the electricity to be of a material nature. When the current-strength or current is an ampère, we say that the unit of quantity of electricity passes in a second; this unit of quantity is called the *coulomb*. The total amount of electricity which has passed through a cross-section of a conductor is obtained by multiplication of the current-strength by the time during which the current has passed.

It is common in the science of electricity to distinguish between electromotive force and potential or tension (potential-difference or tension-difference). The name electromotive force applies to that potential of the element depending upon its chemical composition, and this remains unaltered as long as the element remains constant. It may be compared with the pressure which forces a quantity of water through a pipe. The potential or tension is that electric pressure which we may find at different places along the conductor.

In most courses in physics the following experiment is performed. Water under a certain pressure is driven through a narrow horizontal tube, upon which are a number of perpendicular tubes or water gauges (see Fig. 3).

The height of the water in any of the perpendicular
tubes is a measure of the pressure with which the
water is driven through the horizontal tube at that
point. If we con-
sider the part of
the tube from a to
b, the pressure has
fallen from H to
h, and with the
latter pressure, h,
it makes its exit
from the tube.
The amount of
work which may be obtained when a quantity of
water, M, under the pressure p (per sq. cm.), passes
through the tube is Mp. The quantity of water,
M, in moving from a to b has had its efficiency
lowered from MH to Mh. The quantity of energy,
M (H − h), has therefore been used to overcome the
resistance in the tube, that is, this energy has been
changed into heat which has been absorbed by the
surroundings and consequently lost to us. There
remains only the quantity of work Mh, which is still
available, and may be applied in some way, as, for
instance, in moving a turbine. It is evident how
much depends upon the size of the conducting tube ;
the greater this is chosen the less will be the resistance,
and consequently the greater will be the amount of
available work at the exit.

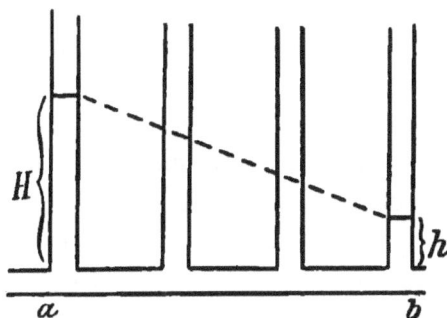

Similar relations exist in the case of the electric
current. Let the wire, AB (Fig. 4), representing a complete electric circuit, be drawn as a straight line. Just
as we measured the pressure of the water in the
tube by its height in the gauge tubes, we may here

FIG. 3.

measure the tension or potential by an electrometer (to be explained later).

We find at A the potential (here electromotive force) π, at B the potential 0, if B is attached to the earth by a conductor. Furthermore, just as previously, when we allow a quantity of electricity (ϵ) to flow through the circuit, we have at A the electrical energy $\pi\epsilon$, and at B, 0. The total energy $\pi\epsilon$ has been changed into heat between A and B, and has disappeared.

If now we cause work to be done, as, for example, in the decomposition of a solution at some point of the circuit, we may use almost the whole of the electrical

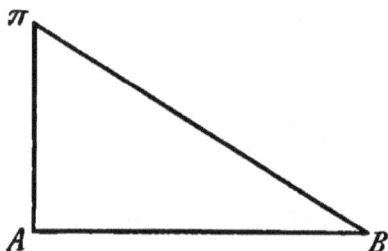

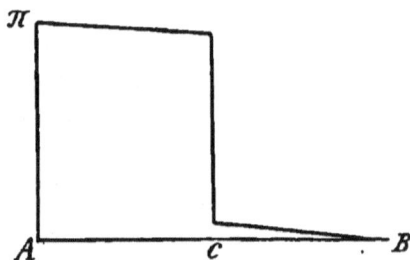

FIG. 4.   FIG. 5.

energy in the work; and moreover, it is immaterial at what point of the circuit we have the work done. Only a very small part of the energy is lost as heat, this amount depending upon the material of the circuit, its sectional area, etc. If we place the solution to be decomposed at c (Fig. 5), an electrometer would show us the above depicted fall of potential at this point, if the quantity of energy $\pi\epsilon$ were almost entirely used for decomposing the solution. Fig. 6 represents the fall of potential in the case where the energy used in doing work is only half of the amount $\pi\epsilon$.

It is possible to use almost all of that energy

which in our analogy with the water was represented by M (H − h), and which was there entirely lost as heat. If we close the tube at b, the pressure rises immediately from h to H, and we obtain at this point the quantity of energy MH, which we may employ as desired. The stream of water differs from the electric current in that the former may leave its conductor while still in possession of a certain amount of kinetic energy, while this property is not possessed by the latter.

We may picture the fall of potential throughout

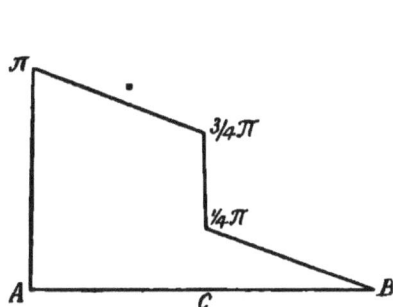

FIG. 6.

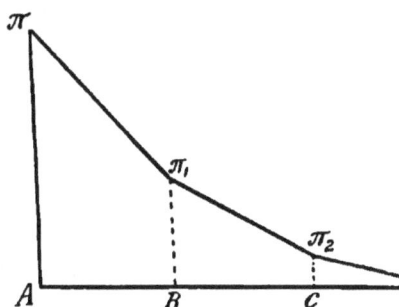

FIG. 7.

any galvanic circuit by the method just employed. At a certain point the potential has its greatest value, and it falls regularly to 0 throughout the circuit when the resistance of the circuit is the same in every part. If work is to be performed requiring a certain amount of electrical energy, and consequently a certain potential, the latter falls by a definite amount at the point where this work is done. If this fall be $= p$, the remaining potential $\pi - p$ falls regularly throughout the whole circuit. If the circuit does not possess the same resistance in every part, the fall of the potential takes place in the different parts in proportion to their

resistance.   For example, in Fig. 7, if the resistance of
AB is twice as great as that of BC and four times as
great as that of CE, the fall of the potential takes
place as represented, π being the electromotive force
in the circuit.   This result follows of necessity from
Ohm's law ($C = \frac{\pi}{R}$), which serves as well for the whole
circuit as for each part.   The value of π for any
portion of the circuit is the difference of potential
between the two ends of that part, and R is the
resistance of the part.   For the case illustrated by
Fig. 7 the following equations are true, since, as also in
the case of the stream of water, the current-strength
is the same throughout the circuit, independent of the
arrangement of the resistances of the parts.

$$C = \frac{\pi}{AB + BC + CE} = \frac{\pi - \pi_1}{AB} = \frac{\pi_1 - \pi_2}{BC} = \frac{\pi_2}{CE}.$$

In consequence of this fact, the potential differences between the single points must be proportional
to the corresponding resistances.   Whether the resistance in the circuit is that of a metallic or of a
liquid conductor, or of both together, this statement
is true.

In a galvanic element whose poles are simply connected by a wire, the resistance of the circuit consists
of that of the wire, called the external resistance, and
that of the liquids which are in the element, its internal resistance.   If the external resistance is 1000 ohms,
and the internal 100 ohms, while the electromotive
force of the element is 1·10 volt, the fall of potential in
the external resistance is 1 volt, and in the internal
resistance 0·10.   Thus we see that there is a difference
between the electromotive force of an element and

the potential - fall which may exist in the circuit outside of the element, and it is evident that the greater the external resistance, the nearer the potential-fall through that resistance approaches the electromotive force of the element. The potential-fall in the circuit is always less than the electromotive force of the element, but approaches the latter as the external resistance approaches infinity, or the internal resistance zero.

We have previously assumed from analogy that the expression $\pi\epsilon$ represents the electrical energy, it being the product of the quantity of the electricity, into its intensity or potential. Could another expression as $\pi\epsilon^2$ .express this energy? This we can determine experimentally. Let us assume that there exists in a circuit an electromotive force, π, expressed in volts, or in other words, the total fall of potential in the circuit is from π to 0. It may be here mentioned that the beginner is inclined to fall into error through the above expression by assuming that the value of π remains the same throughout the circuit, which, as we have seen, is not the case. Let us also assume that in the unit of time the quantity of electricity ϵ expressed in coulombs passes through the cross-section of the conductor, or, as we may also say, since the quantity of electricity in the unit of time is the current-strength, that the current is ϵ expressed in ampères. Imagine the whole circuit placed in a calorimeter. The entire electrical energy is here changed into heat through the resistance of the circuit, and in the unit of time that quantity of heat should be generated which is equivalent to the product $\pi\epsilon$, if this product properly expresses the electrical energy.

If now we employ another circuit in which the

electromotive force is $\frac{\pi}{2}$ and the current-strength 2π, the amount of heat generated in the unit of time would be the same as in the former case, since $\frac{\pi}{2} \cdot 2\epsilon = \pi\epsilon$. In fact, for any values of π and ϵ which give this product, the heat generated in the time-unit would be the same. From experiment we know that such is actually the case. Moreover, if ϵ is kept constant and the electromotive force is made 2π, twice as much heat would be developed as above, and so forth. Consequently it is proven that the product $\pi\epsilon$ is an expression for the electrical energy.

The calculation of the electrical equivalent of heat is now very simple. The unit of electrical energy is naturally the product of 1 volt by 1 coulomb. It is only necessary to measure the heat generated when a coulomb of electricity passes through a circuit whose electromotive force is 1 volt; or expressed differently, when a coulomb experiences a fall of potential of 1 volt, independent of the resistance, since the latter only determines the time in which the fall takes place, while the energy is independent of the time.

If this amount of heat is K calories, $\frac{1}{K}$ is the electrical equivalent of heat, and represents the number of units of electrical energy which are equivalent to the unit of heat. It has been found that volt × coulomb $= 0.236$ cal., or $4.24 \times$ volt × coulomb $= 1$ cal. Moreover, since in mechanical energy 43280 gm. cm. $= 1$ cal., we have volt × coulomb $= 10210$ gm. cm. for the mechanical-electrical equivalent.

$\pi\epsilon$ represents the electrical energy which has passed through a wire between whose ends there was a potential difference, π, and through which the quantity of electricity ϵ flowed. If we allow this energy to change completely into heat, we may write the equation $\pi\epsilon = k$A, where A is the total heat set free, and k a factor depending only upon the ratio existing between the units used in expressing the two kinds of energy. If we represent the current-strength by C, we have πC $= ka$, where a is the heat generated in the unit-time. According to Ohm's law $\pi = k'$CR, and substituting this, we get $C^2R = k''a$, R representing here resistance, and k' and k'' factors of proportion depending upon chosen units. This last equation may be put into the following words : *The heat generated in the whole or a part of a circuit in the unit of time is proportional to the resistance and to the square of the current-strength.* This is known as Joule's law, and was discovered by him in 1841. Its experimental proof is a demonstration of the validity of Ohm's law.

If we choose as units for a, R, and C, the calorie, the ohm and the ampère, then the number of calories generated in the unit of time becomes $0\cdot236 \times$ ampère^2 $\times$ ohm,[1] for volt $\times$ ampère $=$ ampere^2 $\times$ ohm represents the electrical energy available in the unit of time. One such unit, transformed into heat, gives $0\cdot236 \times$ cal. χ units $= 0\cdot236$ χ cal. The number of units (χ) present is expressed by the product ampère^2 $\times$ ohm.

Capacity.—It may be well at this point to explain the term electrical capacity, although it has more to do with statical electricity than with our present subject. It is to be especially noted that this so-called electrical

---

[1] The following may be of interest. A volt-coulomb, called also a joule, is equal to 10^7 ergs, 1 volt-ampère, also called a watt, is equal to $\frac{1}{9\cdot81}$ second-kilogrameter $=\frac{1}{736}$ horse-power $=10^7$ second-ergs, the erg being $\frac{1}{980\cdot6}$ gm. cm.

capacity is quite distinct from the capacity-factor of electrical energy, or the quantity of electricity.

By electrical capacity we understand the ability a body possesses of taking up or holding electricity. This evidently depends upon the nature of the body and also upon the pressure or potential under which the electricity exists. At the same pressure of electric charge the capacities of different bodies are in the same proportion as the quantities of electricity taken up by them. The capacities of bodies upon which equal quantities $_{\prime}$ of electricity are present, under different pressures or potentials, are inversely proportional to those pressures. In general $c = \dfrac{\epsilon}{\pi}$, c being the capacity. The unit of capacity is called the *farad*, and is that of a condenser upon which the quantity of electricity 1 coulomb produces a potential of 1 volt.

Positive and Negative Electricity.—Thus far we have considered the electric current in its analogy to the stream of water, and this is an aid at first to an understanding of the phenomena. The· analogy is not however a perfect one, and care must be taken to prevent misguidance. In the case of the electric current we are dealing with something more complicated than a stream of water.

If we introduce a solution of copper chloride into a circuit as previously described, we observe that while copper is separating at one of the pieces of platinum, chlorine is separating at the other; if we imagine the copper to be transported by the electric current to the one electrode, we must also picture to ourselves the chlorine as carried in the opposite direction to the other electrode. We are obliged then, from this motion of ponderable material in two directions, to ascribe to the electric current, unlike the stream of water, two oppositely directed motions. But we know from the elements of statical electricity, that we have to deal with two kinds of electricity, distinguished by the names positive and negative; hence the conclusion follows that the electric current consists of simultaneous motions of positive electricity in one direction and negative in the other; a conclusion which is supported by electrometric experiments later to be described. The particles of copper always move in the direction of the positive, the chlorine in the direction of the negative electricity.

The conditions found in the case of the factors of electrical energy differ somewhat from those of mechanical energy, as will here be shown. The product volume × pressure represents a quantity of mechanical energy. We know that the capacity-factor, here the volume, is always a positive quantity for we recognise but one kind of volume, but in the electrical energy we have two kinds of capacity-factors, $+ \epsilon$ and $- \epsilon$. For these factors we have the law that the amount $+ \epsilon$ combined with the amount $- \epsilon$ always gives the amount 0. A quantity of positive electricity cannot exist without the existence of the corresponding amount of negative electricity, and the two on coming together neutralise each other. We must accustom ourselves to think of something abstract and cannot expect electrical energy to represent anything as tangible as matter itself. By careful consideration we find, moreover, that if the word substance seems intelligible, we have no reason to consider the expression quantity ⚹ of electricity, or electricity, unintelligible. Let us be clear first as to what we understand by substance. We speak of substance when we recognise a certain number of properties at a single place; one of the properties is, for instance, the occupying of space, that is, the presence of a certain amount of volume-energy. If we compress the substance we diminish its volume, and an amount of work is done which is the equivalent of this compression.

In a similar manner, we learn to speak of electrical energy at a point where we recognise the presence of a number of definite properties or qualities. These properties are not, however, the same as those possessed by matter. A volume-energy cannot be ascribed to electricity, and we cannot grasp it with the hand. It is frequently asked: What are we to understand by a quantity of electricity, or of what nature is electricity? but one seldom inquires of what nature is matter. The former question is really as idle as the latter. The words matter and electricity are nothing more than expressions for a number of definite properties.

We may transform mechanical work into electrical energy, as, for instance, by rubbing a stick of sealing-wax with a woollen cloth, but we always find that the cloth as well as the wax has become electrically charged by the process; the one with positive, and the other with negative electricity. It is a well-known law of nature that whenever electrical energy is produced it always appears in two separate places, though they may lie exceedingly close to each other. We usually speak of a quantity of electricity (ϵ) as passing through the cross-section of a conductor, and also consider it as moving in that direction in which the particles of copper move in the electrolysis; but, as a matter of fact, we must recognise that the quantity $+\frac{\epsilon}{2}$ only is moving in this direction, while

$-\frac{\epsilon}{2}$ always moves in the opposite direction.[1] The movement of positive electricity in one direction is equal to that of the negative in the opposite direction, and we are really therefore justified in considering it as a motion of the two quantities as of one sign, say positive, in the one direction, that of the particles of copper. This is done for simplicity, and we must always bear in mind that it does not represent the exact truth, otherwise we should not be able to understand, for example, the treatment of the following electrometric measurements.

Electrometric Measurements.—In measurements of any kind it is necessary to establish a zero or starting - point. For the intensity - factor of heat - energy, the temperature, we know the absolute zero-point to be $-273°$ C. For the intensity-factor of volume-energy, the pressure, we have a zero-point from which we begin to measure pressures, viz. the pressure zero existing in a vacuum. In the case of speed of motion we do not have such an absolute zero-point, but must always speak of relative motion. In this case we usually consider the motion of the earth as zero, and when we say that a body possesses a speed of v, we actually mean that this is the difference between its absolute rate of motion and that of the earth. We reason very similarly in the case of the intensity-factor of electrical energy, the potential, which only appears in the form of differences, and we know no absolute zero-point from which it may be measured. We take arbitrarily for zero the potential or tension which exists on the earth's surface. If we wish to bring the potential of any point of an electric

---

[1] Exceptions to this rule will be treated under the heading Shares of Transport, or Transference Numbers.

circuit to the potential 0, we simply connect this point with the earth by a good conductor.

Electric potentials are commonly measured by electrometers, of which there are many forms, most of which need not be considered here. The principle is the same in all, and may be understood from a description of one of the simplest forms, the well-known gold leaf electrometer. The two strips of gold leaf hanging together are first connected with the earth, and have then the potential zero. If now, after disconnecting from the earth, we bring into contact with this electrometer a point whose electric potential is to be measured, positive or negative electricity passes from this point to the strips of gold leaf, and these separate, flying farther apart the greater the charge or quantity of electricity given them. This quantity is dependent upon the pressure of the electricity or the potential, and consequently the electrometer is a measure of this potential. The electrometer can be so gauged that the potential, in volts, may be read from a scale attached to it.

Let us now consider for a few moments an electric circuit, the resistance of which is the same in all parts, with a source of electrical energy having a potential of 2 volts at the point AB, Fig. 8. We apply the electrometer to different parts of the circuit, connecting other points with the earth, and thus learn much concerning the nature of the potential in the circuit.

If we connect the middle point of the circuit (C) with the earth, and bring the electrometer in contact with the circuit at A, the source of the positive electricity of the circuit, the electrometer shows us that there is here a potential of one volt, and the electricity may be shown to be positive.

If the electrometer be placed at B, the source of negative electricity, a potential of one volt is also shown, but of negative electricity; at the point C the electrometer shows no potential. Between A and C, and B and C, we find all possible potentials between 0 and 1 volt, the fall of potential being always proportional to the resistance; the only difference being that between A and C the charge given the electrometer is always of positive electricity, while between B and C it is negative. This same arrangement of potential is

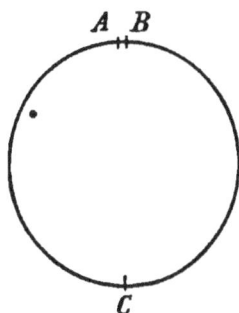

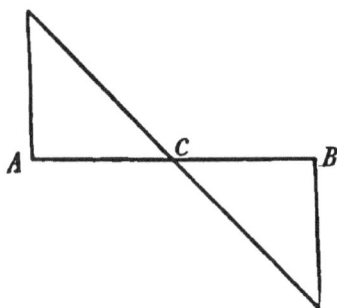

Fig. 8.   Fig. 9.

present if no point of the circuit be connected with the earth, that is, when the circuit is isolated.

If now the circuit be connected with the earth at B, the electrometer shows positive electricity throughout the circuit. At A the potential is 2 volts, at C, 1 volt, and at B, 0 ; between these points the gradual fall is proportional to the resistance.

By connecting A with the earth instead of B, we find only negative electricity in the circuit, 2 volts at B, 1 volt at C, and none at A. The conditions are then comparatively simple, but may be made still more evident by a graphical representation as follows :—
Imagine the circuit unrolled and arranged so that the

line AB is the axis of abscissæ of a co-ordinate system. At separate points throughout the circuit the potential may be represented by ordinates drawn at those points, the potential zero being at the line AB; potentials of positive electricity may then be drawn above, and of negative below this line.

According to this arrangement, Figs. 9, 10, and 11 represent the three cases just considered.

Moreover, we can represent the conditions existing when any point of the circuit is connected with the earth, the potential at that point being zero. We need

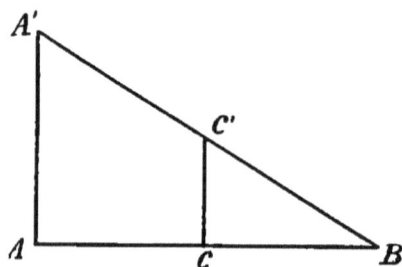

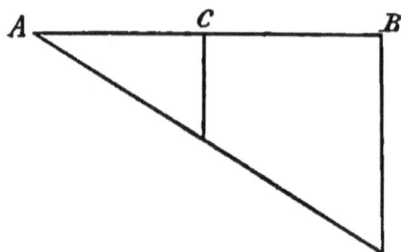

FIG. 10.   FIG. 11.

only to draw through the ordinate of that point, in one of the above figures, a line parallel to the old to serve as a new axis, remove the former, and we have the potentials of positive electricity as before above, and those of negative below the zero line. From Fig. 10, for example, we obtain Fig. 9 by drawing through C^1 a line parallel to AB, and likewise Fig. 11 by drawing the parallel through A^1. In the former case the arrangement is that obtained when C is connected with the earth, C^1 becoming then 0; in the latter case A is connected with the earth, and A^1 becomes 0.

If work is done by the current, and in consequence at a certain point there is a sudden fall in potential, the above method of graphical representation is still as simple as before.

Another property of electrical energy is to be mentioned:—If we have two sources of such energy, as, for instance, two Daniell elements having equal electromotive forces, and we combine the source of negative electricity of each, its negative pole, with the positive pole of the other, the resulting combination has an electromotive force equal to the sum of the forces of the two elements, or 2·20 volts. If they were connected between like poles there would be no current through the circuit. Here we have entirely different relations from those existing in the case of temperature. We cannot add the intensity-factors of heat-energy in this way. If we have two pieces of metal, each having a temperature of 0° C. at one end, while the temperature of the other end is 100°, they can in no way be so combined as to produce a temperature of 200°.

With electrical energy, when a potential difference exists between two points, this difference is not altered through a change involving simply an increase in the absolute potential of those points; and because of this law, we are able to produce an electromotive force of any desired magnitude. If the negative pole of a Daniell element be connected with the earth, the positive pole shows a potential of + 1·10 volt; if now we connect to this positive pole the negative pole of a second Daniell element, we find a potential of + 2·20 volts at the positive pole of the latter, for its negative pole has now the potential of the positive pole of the first element, and the difference between its two poles is 1·10 volt. This arrangement of elements into batteries is commonly called single series or tandem.

Another method, useful for certain purposes, consists in connecting like poles of different elements in
groups, and then connecting by a conductor one of
these groups of like electrodes to the other. In this
way, although no increase in the electromotive force
over that of a single element is obtained, the internal
resistance of the battery thus formed is less than that
of the single element. This is called a parallel arrangement of elements.

No attempt will be made here to give an explanation of the production of the electricity, with its
current in two directions, or of the real cause of the
potential produced at any point. It is sufficient to
comprehend the relations described. To some, however, it is often an aid to consider the analogy of
electrical phenomena to some simpler mechanical
phenomena. The danger of overtaxing the elasticity
of the analogy seems to us to justify an omission of
such simplifying methods.

Having laid the foundations for an understanding
of the electric current, we may now turn towards the
special subject of electrochemistry, and as an introduction to this branch of electrical science, we shall
briefly outline the most important part of the history
of electricity in general. Those desiring fuller knowledge we refer to Ostwald's *Electrochemie, Ihre
Geschichte und Lehre,* and to Wiedemann's *Lehre von
der Electrizität.*

CHAPTER II

A LITTLE more than two thousand years ago the first electrical experiment recorded was performed by Thales, who observed that under certain conditions amber (ἤλεκτρον) possessed the power of attracting light bodies, as pieces of paper, feathers, etc. Later, it was discovered that this property was not confined to amber alone. It was called ἤλεκτρον-like, which later was contracted to the word electric.

The phenomena of atmospheric electricity, as displayed in lightning, St. Elmo's fire, aurora borealis, etc., have always been known, but their consideration as electrical phenomena is of comparatively recent date.

Up to the beginning of the seventeenth century our knowledge of electricity was most limited; at that time it was somewhat augmented by the work of William Gilbert. He showed that a great many substances became electric upon being rubbed, but that none of the metals possess this property. He was the first to declare necessary the rubbing of the material to the production of the electricity.

From this time on, much more interest was taken

in electrical phenomena, and means were soon found for the production of greater electrical effects than were possible through the rubbing of such substances as amber.

In 1733 Dufay first gave expression to the important fact that there are two kinds of electricity, which he distinguished as that produced upon glass, vitreous, and that upon wax, resinous.

At the end of the eighteenth century five different sources of electricity were known. Up to the time of Franklin, friction had been the only source. He observed that the atmosphere was a second source, and a third was found by Wilke in the solidifying of fused substances. This he named "electricitas spontanea." The warming of tourmaline became the fourth source, while the living animal organism offered the fifth when the power of certain fish to produce electric shocks was recognised, as shown by the gymnotus, torpedo, and silurus.

The great electrical discovery of the eighteenth century, the one which attracted the attention of the best investigators of that time, and which has proved to be the discovery of a much more productive source of electricity than was previously known, we owe to the wife of Aloisius Galvani. Galvani was then Professor of Medicine at the University of Bologna. On one occasion he had the freshly-prepared hind legs of a frog lying upon a table, beside which stood an electric machine which was being used. His wife noticed that the frog's legs, which were touching a scalpel, moved as if alive while the sparks were passing from the electric machine. She called Galvani's attention to it, and in a short time he was deeply involved in the study of the phenomenon, considering it a good proof of his theory that the animal organism, in general, was in possession of electricity.

In carrying on his experiments he was accustomed to place the preparations of frogs' legs upon an iron railing in the open air, and he often noticed the contractions taking place in them there, and conceived that it might be due to atmospheric electricity. He found that when lightning was discharged, or storm clouds approached, contraction in the frog's legs was most often produced.

Repeating this experiment during a series of calm, clear days, and observing no effect upon the frog's legs, he twisted the wire which was hooked through the back of the frog, about the iron railing upon which the preparation had been placed, thinking thus more easily to discharge any atmospheric electricity which might have accumulated in the preparation. He observed muscular contractions, which he then concluded were at least not entirely produced by atmospheric electricity. Later experiments carried on in a room showed him conclusively that these same contractions in the frog preparations could be produced without assuming an action of atmospheric electricity, it being only necessary to bring the wire which was hooked through the frog's back into contact with the iron plate upon which the preparation was lying.

The tremendous expansion which the principle involved in this simple discovery received was remarkable. The contractions of the muscles of the frog's legs were recognised as produced by electricity, and the first question arising was as to the source of this electricity.

Galvani declared that the electricity existed in the preparation, which he compared to a charged

Leyden jar. The muscles and nerves replace the coatings of the Leyden jar, and the wire simply serves as the discharging rod. He believed that every animal organism was a source of electricity, this being most evident in the case of the electric eel and certain fishes. He hoped through this discovery to be able to penetrate farther into the mysteries of life in general.

Galvani's opinions were at first pretty commonly accepted by physicists, many of whom repeated the experiments. Even Volta at first was inclined to these views, but later observed that the effects produced were very marked when the material connecting the back of the frog or the nerve with the leg or muscle consisted of two different metals, while the effect was very weak, or entirely wanting, when only a single metal was used. He therefore began to doubt Galvani's explanation, and soon reached the conclusion that the source of the electricity was either in the point of contact of the two different metals forming the "discharging rod," or else at the point of contact between the metal and the liquid (the frog's legs being moist). The preparation itself he considered as nothing more than a delicate electroscope. Volta finally concluded that the principal seat of the electricity was the contact point between the two metals. He believed the action brought about at the point of contact of a metal with a liquid to be of secondary importance. This theory of Volta's has been the commonly accepted theory regarding the source of this electricity until within very recent years.

Volta originally separated conductors into a first and second class—the first comprising the metals, carbon, and certain other good conducting substances occurring in nature, such as the metallic sulphides; the second consisting of all conducting solutions. This distinction is still retained. We describe conductors of the first class as those in which the electric current moves without a simultaneous motion of matter, while conductors of the second class are those in which the transportation of electricity requires a corresponding motion of ponderable material.

Volta soon arranged the conductors of the first class in what was called the electromotive series— that is, he arranged them in such order that if two of them are combined with a conductor of the second class, and also directly with each other, the electric current always passes from the one higher in the series through the liquid to the other. Moreover, the current is greater the farther apart the two chosen metals stand in the series.

After the establishment of this order of electromotive forces of the conductors of the first class, J. W. Ritter made the discovery, entirely unappreciated at the time, that the order is the same as that according to which the metals precipitate one another from solution. Zinc, copper, silver is the order of these three metals in the electromotive series, and zinc precipitates metallic copper from solutions of its salts, and zinc and copper both precipitate silver from its solutions. A connection between electricity and chemistry had thus been practically shown.

A little later Volta stated his law of electromotive forces. This declared that the same potential always exists between two given metals, whether they are directly in contact with each other or form part of a connected series. It explains the impossibility of obtaining an electric current from a circuit made up entirely of metals, for all the electromotive forces which might exist in such a circuit would always give a sum of zero.

The above law, according to Volta, does not hold good for conductors of the second class, because two metals could be connected by a conductor of that class with scarcely any change in the potential from one metal to the other through the solution, since, as he believed, only very slight potential differences were produced at the surface between liquid and metal. Accordingly, the electricity flowing in the circuit

Zinc—Copper

would have nearly the same potential as that between zinc and copper.

Conducting liquid

As long as the attention of investigators was employed with frictional electricity, scarcely any attention was paid to relations which might exist between chemical and electrical processes. Moreover, the quantities of electricity which were produced by the friction methods were too small to bring about any considerable chemical effect. A few experiments were known as early as the middle of the previous century, pointing to relations between these two forms of energy. It was known that by means of the electric spark certain of the metals could be obtained from their oxides; that air, other gases, and water were affected by the passage of the spark had also been observed. The chemical effect of the electric current was first studied on a large scale after Volta had constructed his so-called "electric pile." The latter consisted of pieces of zinc, pieces of pasteboard moistened with a salt solution, and pieces of silver, these being piled in a column in the order given; instead of zinc and silver, other metals could be used. The strength of the pile varied with the choice of the metals and depended upon the number of the pieces from which it was made. Almost every one who was in a position to do so built such a pile, and the scientific papers at the beginning of this century were filled with descriptions of experiments in which the pile was used.

It is worthy of notice that Volta himself says nothing of the chemical action of his pile, in spite of the fact that in his experiments he must have observed the decomposition of water. He evidently could not understand the significance of this phenomenon. The discovery that the voltaic pile could decompose water thus became the work of others.

In the year 1800 Nicholson and Carlisle showed that on conducting the electric current through water, gases appeared at the ends of the conducting wire dipping in the water, one of the gases being hydrogen and the other oxygen, and that except when the wire was of noble metal, it was oxidised. The fact was also not overlooked that the liquid about the wire at which hydrogen was evolved became alkaline, while that about the other wire became acid.

It is surprising that as early as 1802 detailed measurements of potentials of the voltaic pile, which are still accepted as correct, were published by Ermann. Some of the results we have already considered in the introduction; others will now be given. Ermann inserted a silver tube filled with water into the circuit; the ends of the tube were of glass, and through these the wires of a battery were brought in contact with the water. By connecting an electroscope to any desired point of the silver tube, the presence of electricity was shown, and Ermann established the important fact that the column of water between the two ends of the battery wire actually contains electricity during the passage of the galvanic current. The fall of the electroscopic potential, when the column of liquid forms part of the circuit, takes place as we have learned on page 13. In such a case as this, sudden falls of the potential occur at the two poles because of the work performed there.

Ermann also placed wires between the two poles in the tube, as shown in Fig. 12, and observed that gas was evolved at all of the wire ends; in each case

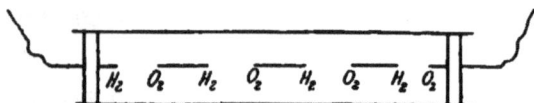

FIG. 12.

an end at which hydrogen appeared was adjacent to one giving off oxygen, as shown in the figure. The conduction of the current occurred in part through the water, and in part through the wires. In this case also the electroscopic potential showed the same arrangement throughout the circuit as before. By properly connecting the circuit with the earth, it is possible to have positive or negative electricity alone in the column of water and the wires; or finally, one part of the pile may be made to exhibit positive, while the rest shows negative electricity.[1]

[1] The discovery first made by Ermann, that when a piece of metal is placed in a liquid through which an electric current is passed, a part of the current goes through the metal, and decomposition of the water takes place at its two ends, has lately received practical application in causing metals to melt under water.

The greater part of the current passing through the metal causes it to become very highly heated, while the water is only moderately heated because of the existence of the well-known Leidenfrost's phenomenon.

The evolution of the gases, hydrogen and oxygen, and the production of alkali and acid at the ends of the wire in the water, was a phenomenon the comprehension of which gave the investigators great trouble. Are these substances produced by the action of electricity upon water ? The law of the conservation of matter was at that time not commonly accepted, so that such a supposition could not of itself be declared absurd, but must first be subjected to experimental proof. Sir Humphry Davy undertook this work, and showed, by very careful investigations, that pure water was decomposed into hydrogen and oxygen by the electric current, but that the formation of acid and alkali was due to impurities. Furthermore, he performed an experiment of the greatest importance upon the migration of acid and alkali to the two poles, for which a satisfactory explanation was not found until after the establishment of the theories of the past few years. This experiment is here briefly described because it relates to phenomena of interest to us. It will be more thoroughly understood after the next chapter has been read. When the reader has become acquainted with the modern theories of electricity, we advise him to attempt to discover an explanation of this experiment. It ought not to be difficult, and he will thereby recognise the advantages of the new conceptions.

If we connect two platinum wires to the poles of a voltaic pile, placing one of the free ends into a vessel filled with pure water, and the other into one containing potassic sulphate solution, the two vessels being connected by means of a tube filled with water, acid is formed at the positive pole (the end of that wire which is attached to the positive pole of the voltaic pile), and alkali accumulates at the negative pole. The same result is obtained if three connected vessels are used instead of two, the electrodes dipping into the end vessels which contain water, while the middle vessel contains the potassic sulphate solution. It looks just as if the positive pole possesses an attraction for the acid, the negative for the alkali, and that in consequence the salt is decomposed.

Davy was seized with the desire to learn more about this motion of the acid and alkali, and by the use of litmus paper he found, much to his astonishment, that the first appearance of acid or alkali was not in the water at the point where it came in contact with the salt solution, but, on the contrary, at the electrodes, whence it gradually spread throughout the liquid. If acid and alkali could pass through the water in going to the poles without affecting the litmus on the way, Davy questioned whether it was not also possible that they might pass through substances for which they had a great affinity without acting upon them, and he found that an interposed acid did not in any way hinder the passage of the alkali to its pole, nor did an interposed alkali solution offer any apparent obstacle to the migration of the acid. There was found, however, in the interposed acid and alkali solutions some of the corresponding salt, just as though the chemical affinity had caused some of the passing compound to be retained. If, when employing potassium sulphate solution, barium chloride solution was used to intercept the sulphuric acid, barium sulphate was formed, and no acid was to be found at the positive pole. Here, thought Davy, the chemical affinity had completely overcome the electrical attraction.

A little later, Davy crowned his experimental work

with the separation of the alkali metals from their solid hydrates by means of the electric current, and afterwards advanced what we may call the first electrochemical theory. This was based upon the atomic hypothesis of Dalton. Experiment had shown that when, for example, copper and sulphur are in contact, the copper becomes positively electric, and the sulphur negatively. It seemed possible, from this fact, that the atoms of two substances when in contact might also in like manner take to themselves charges of electricity. If the electric charges in the atoms were great enough, the differently charged atoms would leave their former positions and come closer together; in other words, a chemical compound would be formed. A decomposition or rearrangement of the atoms would take place if a new atom coming into contact with the previous compound, could assume a charge greater than that already existing on that atom possessing the same kind of electricity; the new atom would attract the atom of opposite sign from its union with the weaker atom, and a new compound would be formed. In agreement also with Berthollet's law of the effect of mass in reaction, he conceived that a large number of atoms with small electric charges might be of greater effect than fewer atoms possessing greater charges.

Davy's theory was not commonly accepted. Berzelius was at that time just beginning his work, and in one of his first investigations, which he undertook with Hisinger, he studied the action of the electric current upon solutions of different inorganic substances, the result of this investigation being the establishment of an electrochemical theory which has been of the greatest importance to chemistry throughout the century.

According to this theory, each atom when in contact with another possesses two poles, one electro-positive, and one electro-negative. When the atoms are in contact one of these poles is usually much stronger than the other, so that the atom acts as if unipolar—that is, electro-positive or electro-negative. The chemical affinity of an element depends upon the amount of the electric charge of its atoms; positively charged atoms react with negatively charged, and the two kinds of electricity partially neutralise each other, the resulting compound being electro-positive or negative, according as the excess of electricity is positive or negative. In this manner the formation of a compound from its elements was explained, as well as the union of two compounds to form a new substance. The existing electric charges of these compounds were thus partially or almost completely neutralised.

An example may make this point clearer. According to the old atomic weights, a positively charged potassium atom combining with a negatively charged oxygen atom resulted in a compound, KO, still possessing a certain charge of positive electricity, as the potassium possessed more positive electricity than the oxygen did negative. A negative sulphur atom combines with three negative oxygen atoms to form the compound SO_3,[1] which is itself negatively charged, because a negative residue results from the union; furthermore, KO and SO_3 combine to form $KOSO_3$, which still possesses some positive electricity. It was supposed that sulphate of alumina,

[1] Berzelius explained the fact of the energetic action between these two negative substances by assuming that the sulphur possessed a comparatively great positive charge as well as its predominant negative charge, and that the negative charge of the oxygen neutralised the former.

$Al_2O_3(SO_3)_3$, was formed in a similar manner, but that it was slightly negative; and the formation of the double salt, $KOSO_3 + Al_2O_3(SO_3)_3$, was therefore explained as the union of the two differently electrical components, sulphate of potassa and sulphate of alumina.

Chemical and electrical processes were closely associated by the above method of reasoning, and the dualistic theory was introduced into inorganic chemistry, or chemistry, since at that time the two were practically synonymous. Every compound was considered as composed of two parts, which might themselves be composed of two other parts. If this theory assumed much that was arbitrary, it performed a great service because of its systematising influence.

From this time no very great advance was made in electrochemistry until Faraday's important discoveries about 1835. Faraday first convinced himself that there was only one kind of positive and negative electricity; that is, whether it was produced by friction or in the voltaic pile, the action was always the same. He then attempted to discover a relation between the quantity of electricity passing through a circuit and the chemical and magnetic effects which it could produce. *He found the three proportional to one another.*

By the comparison of the quantities of different substances which were decomposed by the same quantities of electricity, Faraday discovered a second law, which is proved in the following simple manner. Different electrolytes which are to be investigated are connected into the same circuit in series, so that the same quantity of electricity passes in a given time through each solution. The discovery which he made may be stated as follows: *The quantities of the sub-*

stances, separating at the electrodes in the same time, are in the proportion of their equivalent or combining weights.

If by using platinum poles we connect an acid solution, a solution of a mercurous salt, and one of mercuric salt into the same circuit, and measure the quantity of hydrogen and of mercury, which have separated after a certain time, we find that for every gram of hydrogen liberated in the first solution, 200 grams of mercury are set free in the second and 100 grams in the third. These quantities of mercury are in the ratio of 2 : 1, and correspond to its different valencies in the solutions.

These laws of Faraday have been proved to hold; both that in regard to the proportionality between the quantities of electricity and of decomposed substance, and that concerning the chemical equivalents of the separated substances. It may be stated here that in order to decompose an exact gram-equivalent of any conducting compound, it is necessary to send 96540 coulombs of electricity through the circuit; consequently this number represents the electrochemical unit of electricity; 96540 coulombs will decompose 169·98 grams of silver nitrate. The quantity of silver separated in this case amounts to 107·938 grams. By one coulomb or one ampère in a second, $\frac{107·938}{96540} = 0·001118$ gram of silver are precipitated. We see from these figures that the transportation of very considerable quantities of electricity is brought about by very small quantities of matter.

Faraday's law at first met with great opposition, due principally to the fact that its meaning was not clearly understood. Trouble arose through the misconception of the electrical energy factors; in other words, quantity of electricity and quantity of electrical energy were confounded. The law refers simply to *quantity of electricity,* and asserts the separation of chemical equivalents of substances in the passage of equal quantities of electricity without referring at all to the quantity of *electrical energy* necessary. Among those who made this mistake was Berzelius, who thought that the law required the decomposition of chemical equivalents of all the different electrolytes by the use of equal amounts of energy. This made the law seem absurd, because the chemical affinity or cohesion overcome by the electric current in the decomposition of different compounds is not the same. This mistake and method of reasoning is not entirely of the past, but is still often made.

We have also to thank Faraday for much of our electrochemical nomenclature. Motion of ponderable matter in a solution through which the electric current is passing was assumed to take place in order to explain the observed phenomena. The particles of matter thus moving with the electric current Faraday called *ions,* and gave the name *cathions* to those which move in the same direction as the positive electricity, and the name *anions* to those moving in the opposite direction. Conductors of the second class, or substances which conduct electricity with such an internal motion, he called *electrolytes,* and to the process gave the name *electrolysis.* The name *electrode* he gave to the surface of contact between conductors of the first and second classes. That surface to which the cathions move is the *cathode,* that to which anions move is the *anode.* Although some investigators unfortunately use the word anode to indicate that electrode to which the cathions move, and call the pole to which the anions move the cathode, the terms will be used in this book as they were intended by Faraday.

The Electrolytic Process.—Those who first recognised the decomposition of water by an electric current sought an explanation for the appearance of hydrogen and oxygen at the two electrodes. In the year 1805 Grotthus gave the first complete theory of the phenomenon. According to this theory, in the presence of an electric current, one of the electrodes is positively and the other negatively charged with electricity. The molecule of water (then represented

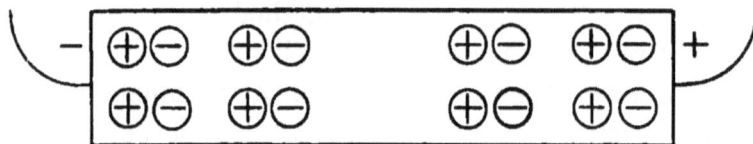

Fig. 13.

by HO) becomes polar—that is, the H is positively electrified and the O negatively. The electrodes attract (and repel) the H and O because of these charges, and the molecules of water between the electrodes arrange themselves as shown in the figure, the positive H being turned towards the negative electrode, the O towards the positive.

If the electromotive force or charges of the electrodes are great enough, the extreme atoms of hydrogen and oxygen are simultaneously liberated. Their electric charges go to neutralise the electricity of the electrodes, and the electrically neutral substances are evolved as gases. The oxygen and hydrogen of these extreme molecules, which are left behind by the liberated hydrogen and oxygen respectively, immediately combine with the hydrogen and oxygen of the adjacent molecules, and this successive decomposition and combination passes on throughout the space between the electrodes. The new molecules then all turn about and again take positions as in the figure, the process proceeding as before. This explanation satisfied the scientific world for many years.

Another point was soon investigated. It was desired to know which really conducts the electricity, the water or the dissolved substance. For a long time this was an open question. One spoke then of "water which by the addition of sulphuric acid, for example, becomes a good conductor," without having apparently conceived any explanation of the fact.

There was also considerable disagreement as to what constituted the positive and what the negative ions of the dissolved substances. Berzelius had brought forward the opinion, at first generally accepted, that for example, in sodium sulphate, which he wrote $NaOSO_3$, NaO was the positive ion and SO_3 the negative, and that these moved to the electrodes, where, by taking up water, they became alkali and acid. Some time later views were expressed to the effect that Na constitutes one ion and SO_4 the other.

These questions were answered by an experiment of Daniell's, which, however, must be considered as decisive only in view of the conceptions then held. He electrolysed sodium sulphate and sulphuric acid solutions separately but simultaneously, in the same circuit, and found that the amounts of hydrogen and oxygen separated were the same for each of the two solutions. He also found that the quantities of base and acid formed in the salt solution were equivalent to the hydrogen and oxygen liberated there; consequently the Berzelius conception must be wrong. According to the latter, by the electrolysis of the salt into base and acid, and the simultaneous liberation of an equivalent amount of hydrogen and oxygen, a double electrical action must have been effected, which is contradicted by the law of Faraday. In agreement with this 'law Daniell declared that the Na must be the positive ion, and the SO_4 the negative; that they give up their electricity at the electrodes, and then act upon the water to produce alkali and acid, and that in this secondary action the hydrogen and oxygen are set free. The amounts of alkali and acid formed must then be equivalent to those of hydrogen and oxygen set free in the solution, and these quantities must also be equal to those of the gases generated from the simple acid solution, as was found to be the case. The salt alone must have conducted the electricity in its solution. The hydrogen and oxygen there evolved are, as shown, the result of a secondary action, otherwise, if the water conducted a part of the electricity, the separated quantities of the gases could not be the equivalent of the acid and alkali produced. In such a case the amount of acid and alkali would be less.

Later experiments of Hittorf and Kohlrausch confirmed Daniell's conclusion; accordingly, the metals and metallic radicles, as H, Na, K, Ag, Hg, Hg^{II}, Fe^{II}, Fe^{III}, NH_4, NH_3CH_3, etc., are positive ions, and the remaining atoms or groups of the conducting substances, as OH, NO_3, Cl, Br, I, SO_4^{II}, $FeCy_6^{III}$, $FeCy_6^{IV}$, etc., form the negative ions. We see here that there are isomeric ions of different valencies, as well among the negative as the positive ions. The trivalent

$FeCy_6^{III}$ is the negative ion of potassium ferricyanide. The tetravalent isomer $FeCy_6^{IV}$ is the corresponding ion of potassium ferrocyanide. The electrical conductivity is a property of the dissolved substance, and not of the solvent.

As the science gradually advanced, the insufficiency of the Grotthus theory began to be perceived. According to this theory, decomposition, and consequently conduction of electricity, could not take place until the electromotive force reached a certain value, below which the affinity or the cohesion of the compounds would not be overcome. But it was found that, with suitable arrangement, the passage of the current took place even when the electromotive force was extremely low. For example, if two silver electrodes be dipped into a silver nitrate solution, a decomposition of the salt can be shown to have taken place, even when the amount of energy used is extremely small. Silver is precipitated upon one electrode, and dissolved from the other, the whole action consisting merely in the passage of silver from one electrode to the other.

According to the Grotthus theory we must imagine the molecule of silver nitrate decomposed at one electrode, and the NO_3 of this molecule recombining with Ag of the next molecule, and so forth to the other electrode, where the last silver atom is oxidised to $AgNO_3$ by the NO_3 set free there. Thus equal numbers of molecules are formed and decomposed. There is here no contradiction of the first law of energetics, but there is of the second law, which may be expressed as follows: *Energy in a condition of rest cannot of itself become active.* To illustrate: a stone lying on the ground cannot of itself rise to a certain height and then fall back again; although such an action would not contradict the first law or that of the conservation of energy, it is contrary to the second. If a stone is to be raised, work must be done upon it from without. The amount of work so employed might be recovered by the fall of the stone back to its original position, but without external aid the stone cannot be raised at the expense of the work to be recovered later.

The Grotthus theory would present us with an exactly similar case. Here the decomposition of the molecules must be brought about by that energy which is recovered when the recombination takes place, and it is this fact that the second law of energetics will not allow. The Grotthus theory requires that the electromotive force shall be above a definite amount before decomposition can take place, which is also, as already explained, contrary to fact. It was Clausius who first showed the disagreement of the theory with the facts. Basing his conclusions upon the experimental material above mentioned, he declared every supposition to be inadmissible which requires the natural condition of the solution of an electrolyte to be one of equilibrium, in which every positive ion is firmly combined with its negative ion, and which at the same time necessitates the action of a definite amount of energy, in order to change this condition of equilibrium into another, differing from it only in the fact that some of the positive ions have combined with other negative ions than those to which they were formerly attached.

The necessary conclusion, from a knowledge of the facts and a consideration of Clausius's statement, is that the individual ions must exist uncombined and free to move in the solution. Clausius was himself prevented from drawing this conclusion by the condition of the chemical theories of his time. In keeping with these views, he attempted an explanation which approaches the present theory. He imagined the positive and negative parts of the molecules in independent motion or vibration, but kept together by their chemical attraction. The latter, he thought, is often overcome by the extreme vibrations, and when the positive part of one molecule comes into a favourable position with respect to the negative of another, these two unite, while their previous companions, momentarily free, come into convenient positions for union with parts of other molecules, and so forth.. In other words, he imagined a continual exchange between the positive and negative parts of the various molecules. When an electric current acts upon a solution, the molecule-parts no longer vibrate and exchange with entire irregularity as before, for decompositions taking place in such a way that they are aided by the electric current, that is, those in which the molecule-parts can follow the direction of the electrical force, become much more frequent than other decompositions. Considering a cross-section at right angles to the direction of the electric current, it is evident that more positive ions would move in the direction of the positive electricity than in the negative direction, and more negative in the negative than in the positive direction. As a result of these different motions, a certain quantity of positive ions passes in one direction, and a quantity of negative ions in the opposite direction. This motion of the two parts of the molecules in the solution causes the conduction of the electricity. According to this theory of Clausius, the current does not cause

any decomposition of molecules, but only guides those
molecule-parts which are momentarily free, so that
their motion is in the direction of one of the oppositely
charged electrodes.   This theory was very commonly
accepted, and has been almost until the present time.

At about the same time that Clausius brought
forward his ideas, Hittorf began his work upon the
migration of the ions, and a little later Kohlrausch
commenced his experiments upon the conductivity
of solutions.   Through these investigators a great
advance was made, and from their acquisitions,
Arrhenius in 1887 replaced the Clausius theory by
the theory of free ions.

**Relation between Chemical and Electrical
Energy.**—Before closing this brief historical account
it is necessary to state that, soon after the establishment of the law of the conservation of energy, attempts
were made to answer the question: Does the chemical
energy of the process taking place in a voltaic element,
as measured by the heat generated, change completely
into electrical energy?

The Daniell element consists of zinc, zinc sulphate,
copper sulphate, and copper, and when in action, zinc
goes into solution while copper separates out.   The
generation of heat corresponding to this reaction is
known from thermochemical measurements; for the
gram-equivalent of the two substances it amounts to
25050 cal.

$$CuSO_4 + Zn = ZnSO_4 + Cu + 25050 \text{ cal.}$$

If this reaction yielded only electrical energy, the
electrical equivalent of 25050 cal. would be produced.   On the other hand, the amount of electrical
energy actually obtained from the element may be

E

easily calculated. The above reaction represents the case when one gram-equivalent of copper has separated, consequently 96540 coulombs of electricity have passed through the circuit. For it follows from Faraday's law that this amount of electricity always passes through the circuit when the electric deposition or solution of a gram-equivalent of any substance takes place.

The electromotive force π of the element can be measured, and it is known that volt $\times$ coulomb $= 0.236$ cal.; consequently in order to express the electric units volt $\times$ coulomb in calories, it is necessary to multiply their number by 0.236.

The electrical energy expressed in calories is therefore

$$0.236 \times 96540 \times \pi \text{ cal.}$$

The chemical energy in heat-units is 25050 cal. If they are equal in this case,

$$0.236 \times 96540 \times \pi \text{ cal} = 25050$$

and

$$\pi = \frac{25050}{22784} = 1.10 \text{ volt.}$$

This calculated value 1.10 volt is also the electromotive force of the Daniell element experimentally found, and it was concluded from the agreement in this case that the chemical energy of a reaction is changed completely into electrical when that reaction is the source of the electric current from an element.

Later experiments with other elements gave results not entirely agreeing with this conclusion. The question was finally answered by the theoretical and experimental investigations of Willard Gibbs, F. Braun, and H. von Helmholtz, who showed that there is usually a difference in the amounts of chemical energy transformed in an element and electrical energy obtained therefrom. This difference is manifested by a generation or absorption of heat in the element when it is in action.

CHAPTER III

THE ARRHENIUS THEORY OF DISSOCIATION

ELECTRICAL investigation received a great impetus from the theory of Arrhenius [1] in 1887. Well-known facts whose relation to one another was previously unknown became connected by this theory, and it was a great impetus to new discovery. The scientific electrochemistry of to-day has this theory for its foundation. We shall consider in detail its development, and shall then ascertain the present position of electrochemistry as it appears in the light of this new conception.

In 1887 J. H. van't Hoff published an article in the first volume of the *Zeitschrift für physikalische Chemie* upon the *rôle* of osmotic pressure in the analogy between solutions and gases. He had established theoretically and experimentally the following very important generalisation of Avogadro's law.

"*At the same osmotic pressure and temperature, equal volumes of all solutions contain the same number of molecules, and, in fact, that number which under the same pressure and at the same temperature exists in the same volume of a gas.*"

[1] *Zeitschr. physik. Chem.* i. 631, 1887.

It was likewise shown that the gas laws of Boyle and Gay-Lussac applied also to dilute solutions.

What is to be understood by osmotic pressure may be made clear by the following experiment. A vessel is filled with water, and in it a vertical tube, closed at its lower end by a semi-permeable membrane and open at the upper end, is placed. A quantity of some solution, for example, of sugar, is poured into the tube until the heights of the liquids outside and inside are the same. The semi-permeable membrane here used is of such a nature that the water may pass through it while the dissolved sugar is prevented from doing so; such membranes are not difficult to prepare. It is observed in this experiment that the column of liquid in the tube begins to rise, water entering from the outer vessel through the membrane. A certain pressure must be exerted upon the liquid in the tube in order to prevent its rising. That pressure, which will just hold the level of the liquid in the tube in its original position, is the equivalent of the osmotic pressure. This osmotic pressure of dissolved substances corresponds to the pressure of gases.

It is known that the equation $pv = \mathrm{RT}$ expresses Avogadro's, Boyle's, and Gay-Lussac's laws regarding gases, v being the volume in cubic centimeters of a gram-molecule of the gas under the pressure p expressed in grams per square centimeter, T the absolute temperature, and R a constant. The expression $\frac{pv}{\mathrm{T}}$ has a constant value for a perfect gas, independent of its nature and condition of dilution. This value is represented by R. The expression is the result of experimentally discovered facts, though not obtained in a direct manner. Whenever the molecular volume of a gas is multiplied by its pressure and the product divided by its absolute temperature the constant R is obtained, which, when the units are those above described, is 84700.

This gas equation also holds good for dissolved substances. Pfeffer found the osmotic pressure of a 1 per cent sugar solution at 6·8° C. to be equal to 50·5 cm. of mercury or 50·5 × 13·59 gms. There being nearly one gram of sugar in 100 cubic centimeters of the solution, the molecular weight in grams (342) is contained in 34200 cubic centimeters, and this is then the value of v or the molecular volume. T was in this case 279·8. Consequently for this sugar solution

$$\frac{pv}{T} = \frac{50\cdot5 \times 13\cdot59 \times 34200}{279\cdot8} = 83900 \text{ (approximately)}.$$

This value of R only differs from the value for gases by the possible errors of experiment. Evidently then the osmotic pressure of the sugar solution is the same pressure as the sugar would exert if it existed in the gaseous state and occupied the same volume.

Because of the prominence of osmotic pressure in the considerations of the following pages, it is well at this point to obtain an idea as to how it is produced. If the lower end of the tube in which the sugar solution is placed be entirely closed, we of course observe none of the evidences of this pressure. At the limiting surfaces of a solution there exists a pressure—the internal pressure—acting inward at right angles to the surface and amounting to over a thousand atmospheres.[1] In a 1 per cent

---

[1] We are obliged to recognise the existence of such a pressure by certain experimental facts which cannot be here described.

sugar solution there is an osmotic pressure of only about one atmosphere directed against this enormous internal pressure. This pressure is due to the sugar, which acts in the water just as it would if it were in the gaseous state and confined in the same space. Even with very concentrated solutions the internal pressure is still hundreds of atmospheres greater than the opposite or osmotic pressure exerted by the dissolved substance. It is on this account that the vessel containing a solution is not broken by the osmotic pressure exerted against its walls. Besides the weight of the solution itself there is no pressure acting upon the containing vessels.

By employment of the semi-permeable membrane we are enabled to observe the effect of the osmotic pressure. When the tube is closed at its lower end by this membrane, and placed in water, the water enters through the membrane. In the solution, at all surfaces, the internal pressure A of the water is exerted inward and the osmotic pressure b of the dissolved sugar outward, while in pure water only the pressure A exists. Upon the membrane only the pressure b is exerted, because the membrane is permeable to the water. There being at the membrane no liquid surface, there is consequently no manifestation of internal pressure. Leaving out of account the internal pressure of the water, there is exerted upon the surfaces of the solution and the membrane the pressure due to the dissolved sugar, which of course disappears when pure water is used to replace the solution. The solution therefore tends to expand when in contact with the water of the outer vessel, and can do so at the expense of this water, which enters through the membrane.

It is evident that the membrane enables us to
observe the existence of osmotic pressure, and this
may be defined as the pressure exerted on the membrane.　The rising of the water in the tube may be
more easily understood by calling to mind an experiment with the air-pump.　If water be placed in the
tube with the permeable wall and in the outer vessel
as well, it could be made to rise in the tube by
diminishing the atmospheric pressure acting upon it
there.　In this way we diminish the external pressure
which is acting inward.　By dissolving sugar in the
water, a pressure is created inside the solution directed
outward.　The result will evidently be the same,
whether the former be the case or the latter.　The
liquid must rise in the tube, as it actually does.　In
order then to explain the osmotic pressure, we evidently
need not make any assumption as to an attraction
existing between the solvent and the solution, but
only the assumption that the substance in solution
acts as it would in the gaseous state.

Van't Hoff has, in fact, proved the existence of
far-reaching analogies between dilute solutions and
gases, and has also been able to deduce laws for phenomena not apparently related to osmotic pressure,
from the laws of osmotic pressure itself.　Among such
phenomena may be mentioned the influence of a dissolved substance upon the vapour pressure and upon
the freezing point of the solvent.　These laws had
already been discovered, principally by Raoult, and
were thus expressed : *the lowering of the freezing point
and vapour pressure of a solvent by a dissolved substance
is proportional to the concentration, and is the same for
equi-molecular solutions*—that is, such as have, in equal
amounts of the solvent, quantities of the dissolved

substances proportional to their molecular weights. These discoveries furnished an opportunity of making great advances in our knowledge of the conditions of matter, and especially gave us a simple method of determining the molecular weights of all soluble substances, while this could previously be done only for those which are volatile.

One great difficulty presented itself, and cast a dark shadow upon the otherwise bright theory of solutions. Almost all acids, bases, and salts which were soluble in water gave agreeing results for molecular weights by the methods of the osmotic pressure, freezing point, and vapour pressure lowering, which were much lower than those obtained by the vapour density method, and were less than expected from the chemical properties of the substances. In other words, assuming the normal molecular weights, these showed too great osmotic pressures and freezing point depressions.

Not very long before, the molecular theory had been in a very similar position, because of the deviations of many vapour densities from the requirements of the theory. It was only with considerable hesitancy that the explanation was admitted that there was a dissociation of the molecules of these gases. At the present time this hypothesis is accepted. It seemed natural to expect a similar dissociation in solutions, and this assumption was first made by Planck,[1] who based his reasoning upon thermodynamical considerations. This explanation of the difficulty was not then accepted by chemists. Such a supposition seemed absurd, for it required that substances like potassium chloride, in which the attraction holding the atoms together was considered very great, should decompose

[1] *Zeitschr. physik. Chem.* i. 577, 1887.

into potassium and chlorine, and that these should exist as such in the solution, in spite of the fact that potassium reacts so energetically with water. The supposition also seemed to be contradicted by the law of the conservation of energy, for the assumption implied that substances which combined so energetically that much heat was generated, should separate again, apparently of themselves.

Before such an essential change could be made in the opinions held regarding solutions the apparent contradictions had to be removed. This was done by Arrhenius, who was able not only to do away with the seeming contradiction, but also to produce clear evidence for the truth of Planck's hypothesis. In an early investigation of the electric conductivity of solutions, Arrhenius had recognised two kinds of molecules, and had concluded that only one of these kinds, the active molecule, caused the conductivity and that the other kind was inactive. He also expressed the opinion that all inactive molecules changed into active in solutions of extreme dilution. He recognised an activity coefficient of a solution, this being the relation between the number of active molecules and the sum of active and inactive molecules therein, and that at infinite dilution this coefficient would be equal to unity. For other dilutions it was less than unity, and expressed the relation between the existing molecular conductivity, of which we shall soon learn more, and its limiting value, or the molecular conductivity at infinite dilution. It was unknown in what respects the active molecules differed from the inactive. As soon, however, as the above-mentioned work of van't Hoff appeared, Arrhenius was able, by comparing the effects of the electrolytes in depression of the freezing point of water with their electrical conductivity in solution, to adduce perfectly convincing proof of electrolytic dissociation. This proof was published in the article previously referred to, and entitled, " Über die Dissociation der im Wasser gelösten Stoffe."

As already remarked, there is a class of compounds, such as sodium chloride, which give too great a reduction of the freezing point. A gram-molecule of sugar dissolved in ten liters of water produces a reduction of the freezing point of about $0.186°$, while a gram-molecule of sodium chloride gives nearly twice that reduction. It is evident, if we accept van't Hoff's assumption applied to this case, and consider the molecules of salt as dissociated into sodium and chlorine atoms, the extent of this dissociation may be calculated from a knowledge of the deviation of the freezing point depression from the value for the undissociated substance. If i represents the ratio between the actual depression of the freezing point and the depression which the substance would give if normal or undissociated, and k is the number of parts into which each molecule divides (2 in the case of NaCl, for $MgCl_2$ 3, etc.), while a stands for the degree of dissociation, that is, the number of dissociated molecules, divided by the total number of molecules, then :

$$i = 1 + (k - 1)a,$$

and

$$a = \frac{i - 1}{k - 1}.$$

This degree of dissociation represented by a, called by Arrhenius the affinity coefficient, was calculated by him for a great many substances from their known freezing point depressions, and it was found that these results agreed with the dissociation values which the electrical conductivity had given. Only those substances conduct which are at least partly dissociated, and therefore the conductivity is due to the dissociated parts; to the latter, which were called by him the "ions," Arrhenius ascribed electric charges. He did not fail to call attention also at that time to the fact that many other physical and chemical phenomena receive a great deal of light from this recognition of the free ions.

Evidently we have here a dissociation differing from that which ammonium chloride undergoes when heated. The parts resulting from the decomposition are electrically charged, and contain equivalent amounts of positive and negative electricity. It is natural to ask: Whence come these sudden charges of electricity? They seem to be produced from nothing. An answer which seems satisfactory is not difficult to give. It is known that metallic potassium and iodine combine to form potassium iodide. In this combination heat is generated, which shows that the two have entered a state in which they contain less energy than before. A certain amount of chemical energy doubtless still remains in the compound, and when the salt is dissolved in water, the greater part of this chemical energy is changed into electrical, through the influence of the solvent. This is the energy seated in the charges of the ions. The potassium ion is positively, and the chlorine negatively electric. By aid of the electric current, it is possible to add to these ions the energy in the form of electricity necessary to give them the energy they originally possessed as elements. In such a case they separate in the ordinary molecular forms at the electrodes.

From this consideration the difference between ions and atoms, or molecules of an element, is made clear. They contain usually very different quantities of energy. Elements in their natural or molecular state differ so widely from the ions in all their properties, that we may say they have nothing to do with each other further than that one may change into the other.

Although the theory of dissociation in solution encountered a great many opponents in its early years, it has nevertheless successfully advanced, and to-day there are few who openly oppose it. The benefits being constantly derived from it are very great, and we shall be continually reminded of its value and utility.

CHAPTER IV

THE MIGRATION OF THE IONS

In the aqueous solution of an electrolyte we recognise the existence of free ions, each possessing a definite electrical charge. For example, in the solution of hydrochloric acid there are the hydrogen ions charged with positive and the chlorine ions charged with negative electricity. We can now express Faraday's law by saying first that conduction of electricity through a solution is only brought about by the movement of those ponderable particles which are charged with electricity, in this case the hydrogen and chlorine ions, and secondly, that chemically equivalent quantities of the substances are charged with equal amounts of electricity.

A galvanic or an electric current may be produced in an electrolyte by dipping into it two electrodes, which are connected, one with the positive and the other with the negative pole of a source of electricity. In consequence of the difference of potential thus produced, motion of the ions is brought about, and an electric current passes through the solution. Under such circumstances a decomposition of the electrolyte always takes place, even though this may not be manifest. With hydrochloric acid, gaseous hydrogen and chlorine separate in unelectric form at the electrodes. An electric current can also be produced by induction without the use of electrodes, in which case no transformation of the conducting particles from the electric (or ionic) into the unelectric or molecular state takes place.

When an electric current is conducted through a solution, a certain number of positive ions pass through a cross-section of the solution in one direction, and simultaneously a certain number of negative ions in the opposite direction; and it was previously believed that when the two were of the same valency this rate of motion of the positive and negative ions was the same for both, undoubtedly because of the fact that equivalent quantities separate at the electrodes in a given time. We know now, however, that these rates of motion are seldom the same for different ions, for the phenomena of electrical conduction and precipitation are not so closely connected as at first seemed probable. Their relation, and the general division of the work of conductivity among the different possible ions, were discovered by Hittorf.[1] This knowledge was obtained by a careful study of the changes in concentration taking place about electrodes when the electric current is being passed through solutions.

It will now be seen how it is possible to learn from these concentration-changes the relative rates of migration or motion of the ions.

Migration of the ions takes place whenever a solution of an electrolyte conducts electricity, and the ions separate at the electrodes in the molecular condition.

---

[1] *Pogg. Ann.* 89, 98, 103, 106 (1853-1859). Collected in Ostwald's *Klassiker d. exakt. Wiss.* Nos. 21 and 23.

Since now there are always present in the electrolyte equivalent amounts of positive and negative ions, it follows that, if at the positive electrode a negative ion separated without the simultaneous removal from the solution of a positive ion at the other electrode, the solution would contain more positive than negative ions. In other words, it would be charged with positive electricity, and because of the great quantity of electricity which characterises an ion, the charge would be strong. If still another negative ion were to separate alone, a greater amount of work would be necessary than before, because the solution, being positively charged, would now have a greater attraction for the negative ion, and resist its separation. On the other hand, the separation of a positive ion at its electrode would become easier than before, because of the repelling force of the positive electricity of the solution. Since this electrostatic force, compared to the others coming into play, is very great, the decomposition of the electrolyte must take place in such a manner that the positive and negative ions always leave the solution at such rates that the solution itself remains electrically neutral.

Besides this fact of the simultaneous separation of equivalent positive and negative ions, it is also known that the current-strength, or the quantity of electricity passing through a cross-section of the electrolyte in the unit time, is the same at all points of the circuit. Now the total quantity of electricity in motion is evidently the sum of the two oppositely-directed quantities of positive and negative ; but there is no reason why, if in one part of the circuit the quantity of electricity 1 consists of $\frac{1}{2}$ positive and $\frac{1}{2}$ negative, it may not at another point be composed of $\frac{1}{4}$ positive

and $\frac{3}{4}$ negative electricity ; for the motion of the one
kind of electricity in one direction is equivalent to
the motion of the other kind in the opposite direction.
Consequently we are justified in considering the current
as a motion of the one kind of electricity in a single
direction.   As a matter of fact, any portion of the whole
electricity in motion may move in one direction, while
the remainder is oppositely directed.   In consequence
of these statements there is no necessity for assuming
equal rates of speed for the different ions.   This would
only be the case when there was a motion of equal
quantities of positive and negative electricity at the
same rate in opposite directions.

In metallic conductors, or conductors of the first
class, equal quantities of positive and negative electricity flow in the same time, but this is seldom true of
the conductors of the second class, the electrolytes.
This arises from the different degrees of mobility
possessed by the ions ; the mobility of the chlorine
ion, for example, differs very much from that of the
hydrogen ion.   When they are subjected to the same
force, as in the electrolysis of a solution of hydrochloric acid, the hydrogen moves, in fact, about five
times as fast as the chlorine.   It will presently be
seen that a number of facts concerning conductivity
in solutions may be explained by the assumption of
different rates of motion for the different ions, but it
must always be remembered that the same quantity
of positive electricity in cathions must be present in
any part of a solution as there is negative in anions
there.

The motion of the different ions may be illustrated
by comparing them to two companies of cavalry going
in opposite directions.   Suppose one company walking,

F

the other galloping, and imagine a ditch in the way, which they all cross. If the second company moved five times as fast as the first, five horsemen would cross the ditch in one direction, while one was crossing in the other, or, of the six crossing the ditch in a given time, five belong to the second, and one to the first company. If each of the horsemen carried a sixth of a bushel of oats with him, one bushel would cross the ditch in the unit time, though $\frac{5}{6}$ of it go in one direction and $\frac{1}{6}$ in the other. The oats here represent the electricity.

When a solution of hydrochloric acid forms part of an electric circuit, the conduction of a quantity of electricity through the solution takes place in such a manner that $\frac{5}{6}$ of it, as positive electricity, moves in one direction with the hydrogen ions, and $\frac{1}{6}$ in the other with the chlorine.

The effect of these different rates of migration upon the composition or concentration of the different parts of the solution may be easily calculated: Suppose a solution of hydrochloric acid containing 30 gram-equivalents be placed in a vessel between the electrodes A and B (Fig. 14). In each third of the vessel there are then 10 gram-equivalents. If 96540 coulombs be conducted through the solution, one gram-equivalent of hydrogen and one of chlorine will separate at the electrodes A and B, and we may imagine these gases removed from the solution. The amount of electricity, 96540 coulombs, must have

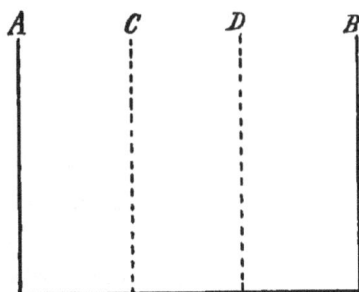

Fig. 14.

passed through every cross-section of the solution, therefore through C and D.

If both ions migrate with the same speed, one half of a gram-equivalent of H ions, carrying 48270 coulombs, has passed from BD through DC to AC, and half a gram-equivalent of Cl, also carrying 48270 coulombs, has moved through DC from AC to DB, or one gram-equivalent of ions has passed the cross-sections C and D. One gram-equivalent of hydrogen has been removed from AC as gas, while, according to the supposition, one half a gram-equivalent has come in from DC, therefore AC now contains $9\frac{1}{2}$ gram-equivalents of H ions. Since one half of a gram-equivalent of Cl has passed from AC, there are also $9\frac{1}{2}$ gram-equivalents of Cl there. Similarly DB contains $9\frac{1}{2}$ gram-equivalents of H and Cl. It follows, then, that when the rates of migration of the ions are the same, the relation of the concentration in AC and BD remains unchanged. The solution in the middle division, DC, has the same concentration as originally, that is, 10 equivalents, because the same number of ions have entered this portion as have left it.

The hydrogen ions really migrate about five times as fast as the chlorine, and the above consideration must be altered accordingly. There have actually $\frac{5}{6}$ gram-equivalents of H ions with $\frac{5}{6}$. 96540 coulombs passed from BD through DC to AC, while $\frac{1}{6}$ gram-equivalent of Cl ions with $\frac{1}{6}$. 96540 coulombs have passed from DB to AC, or, in all, one gram-equivalent with 96540 coulombs has passed through the sections C and D. The composition of the middle portion remains unchanged as before; AC and BD have undergone the following changes: one gram-equivalent of H ions has left the solution in AC, being evolved as gas, $\frac{5}{6}$ of an equivalent have entered AC, so there are now here $9\frac{5}{6}$ gram-equivalents of H ions; there is also the same number of Cl ions in this portion of the solution, because only $\frac{1}{6}$ gram-equivalent of Cl has left it. In BD there are left only $9\frac{1}{6}$ gram-equivalents of H ions, since $\frac{5}{6}$ have passed from this portion to AC, and there are also $9\frac{1}{6}$ gram-equivalents of Cl here, for one equivalent has been evolved as gas at the electrode, and only $\frac{1}{6}$ equivalent has entered from CD. There are then $9\frac{5}{6}$ gram-equivalents of hydrochloric acid in AC, and only $9\frac{1}{6}$ in BD, or AC has suffered a loss of $\frac{1}{6}$, while BD has lost $\frac{5}{6}$ equivalents.

From these two examples the rule is derived that the loss at the cathode (in AC) stands in the same ratio to the loss at the anode (BD) as the rate of migration of the anion (Cl) to that of the cathion (H) ($1 : 5$ in this case).

It was in the manner just indicated that Hittorf was able to determine the relative rates of migration of the different ions from the changes taking place in the concentration of the solution near the electrodes. His conclusions, though they at first met with some opposition, are now generally accepted.

From a superficial consideration of the question, one is inclined to believe that when one of the ions migrates faster than the other, positive ions must accumulate in one part of the solution, and negative in another.   That this is not the case has, however, already been observed.   If, for example, x be the amount of positive ions separated from the solution, and y the amount which has come to AC from BD through CD, then AC has $x - y$ less positive ions than before the passage of the current.   The same amount

$x-y$ of negative ions must in this case have gone to
BD, for if x ions have separated at the electrode, x
ions must also have passed the sections C and D, and
y have gone from BD to AC. Similar considerations
apply to BD. In order that the law of electricity—
that the current-strength shall be the same in all parts
of the circuit—may obtain, the relations must be as
here described.

A second question which might naturally present
itself is: How can one gram-equivalent of chlorine
separate at the electrode B, when only $\frac{1}{6}$ gram-equivalent has passed through any section of the solution ?
To explain this it is assumed that there is always a
large excess of ions in the immediate vicinity of the
electrode, so that in any given time more may
separate at the electrode than migrate towards it.
The phenomenon of ordinary diffusion assists in this
case.

The determination of the ratio of the rates of migration of two ions is as simple as it appears from the
above. It is only necessary to divide the solution,
whose concentration is known, into three parts, and,
after the passage of a known quantity of electricity, to
measure the concentration-changes which have taken
place. The middle portion must always remain unaltered. This being the case, the material which has
been concentrating at the electrode has not diffused
from that portion of the solution, and thus destroyed
the value of the results.

We will represent by 1, expressed in gram-equivalents, the amount of cathion or anion (since the two
quantities are always alike) separated at the electrodes,
and by n that portion of a gram-equivalent of cathion
which has passed from the anode to the cathode ; then

$1 - n$ gram-equivalents of anion must have migrated from the cathode to the anode. These quantities, n and $1 - n$, are called the shares of transport or transference numbers of the cathion and anion respectively, and their ratio gives us, as above described, the ratio of the velocities of migration.

$$\frac{n}{1 - n} = \frac{u}{v} = \frac{\text{loss at the anode}}{\text{loss at the cathode}},$$

where u represents the velocity of migration of the cathion, and v that of the anion. Finally, the relative rate of migration of an ion is the quotient obtained by dividing the distance over which it has gone by the sum of the distances both ions have covered. For the cathion this would evidently be represented by n, and for the anion by $1 - n$, since the distances are proportional to the transported amounts.

As is evident, the relative rates of migration of the ions can be experimentally determined, but their actual velocities expressed in a definite unit of measurement cannot be learned in this way.

The following example illustrating the method employed is taken from Hittorf's work. It will make the point still clearer, and show how the calculation is easiest made.

A solution of silver nitrate was electrolysed for some time, and the quantity of precipitated silver determined. This amounted to 1·2591 gram. A certain volume of the solution about the cathode gave 17·4624 grams of AgCl before, and 16·6796 after the electrolysis. It had lost 0·7828 gram of AgCl, or 0·5893 gram Ag. If no silver had come into this portion of the solution, its loss would have been the silver precipitated on the electrode, but being found only 0·5893

gram poorer in silver, $1\cdot2591 - 0\cdot5893$ or $0\cdot6698$
gram of silver must have migrated into it. If as
much silver had migrated as was precipitated ($1\cdot2591$
gram), the share of transport for silver would have
been $= 1$, and the NO_3 ion would not have taken part
in the migration. But only $0\cdot6698$ gram of silver
actually migrated, so the share of transport for silver is
$\dfrac{0\cdot6698}{1\cdot2591} = 0\cdot532$. For the NO_3 ion it is $1 - 0\cdot532 =$
$0\cdot468$. The solution about the anode might have been
analysed to serve as a check for the above. A loss of
$0\cdot6698$ gram of silver would have been recognised at
that point.

When it is preferable from an analytical standpoint,
the analysis may, of course, be carried out for the
anion instead of the cathion, if, for the sake of greater
certainty, both anion and cathion are not investigated.

An example of this is found in the determination
of the shares of transport for cadmium and chlorine.
In this case the anode consisted of amalgamated
cadmium, which formed cadmium chloride with the
chlorine liberated. From the loss in weight of the
anode the quantity of chlorine separated was determined. The original concentration of chlorine at the
anode was known, and after the electrolysis it was
again determined. All of the cadmium chloride formed
remained in this portion of the solution, and the corresponding amount of chlorine was subtracted from
the total amount found here. This remainder, when
subtracted from the original amount present, gave the
" loss," from which, as before, the transported quantity
of chlorine was calculated. This is the difference
between the amount of chlorine which combined with
cadmium and the above " loss."

There are a great many forms of apparatus which have been used for the measurement of these quantities. In order to give an idea of the essential features of such an apparatus, one used by Nernst and Loeb for determining the shares of transport for silver salts is represented by the cut (Fig. 15). The two electrodes are silver. Upon the cathode a quantity of silver was precipitated, which was in each case a measure of the quantity of electricity which had passed. The same amount of silver was simultaneously dissolved from the anode. The apparatus resembles the Gay-Lussac burette in form. In order to do away with the disturbance caused by the falling of silver from the cathode, this was placed in the side tube, being introduced through the small tube $\overset{\circ}{B}$ into the bulb at its bottom. This electrode consisted of a cylindrical piece of silver foil attached to a silver wire.

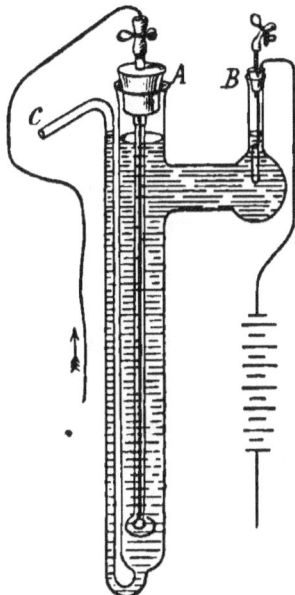

FIG. 15.

The anode, which was a silver wire twisted at its lower end into a spiral, was introduced through A, and reached to the bottom of the longer tube. The straight portion of this electrode was encased in a glass capillary. At A and B were corks, each pierced by a short piece of small glass tubing. The piece in A simply allowed of the passage of the electrode wire, while that at B had a piece of platinum wire fused into its side, upon which the electrode could be hung. With this

arrangement it was possible to remove portions of the
solution from the apparatus without disturbing the
electrodes, by drawing or forcing the liquid through
the rubber tubes which are shown connected to the
glass tubes at A and B.

In performing an experiment with such an apparatus the electrodes and corks were weighed. When
the apparatus had been put together, the opening at A
was closed, the opening of the tube C was placed in
the solution to be used, and by sucking at B, the whole
was filled to a point above the side tube. The tubes
usually held from 40 to 60 cubic centimeters of solution. The exit tube was then closed with a rubber
cap, and the whole placed in an upright position in an
Ostwald thermostat. After the temperature was properly
adjusted the current was conducted through the solution.
Immediately after disconnecting the current the exit
tube was opened, and by blowing at the opening B, the
desired quantity of solution removed into a properly
tared vessel. It was then weighed and analysed.
The amount of solution remaining in the tubes was
determined by weighing the whole and subtracting
the weight of the apparatus. In case no considerable
motion had taken place within the solution, as through
diffusion or convection currents, the altered portion of
the solution about the anode was removed in the first
small portion, and there was sufficient of the unaltered
middle portion to serve for washing out from the
electrode the solution of changed concentration. The
following portions of solution were then unaltered in
concentration, while that part of the solution which
was about the cathode remained in the apparatus. A
test of the accuracy of the experiment was found in
the unaltered condition of the middle portions of the

solution, as well as in the fact that the solution about the cathode had lost as much silver as that at the anode had gained.

Hittorf asked himself at the beginning of his work, Are these shares of transport constant, or are they variable under varied conditions? and, if they are variable, upon what do the variations depend?

He recognised three points to be taken into consideration—the influence of the current-strength, that of the concentration, and that of the temperature. He found that the velocities of migration of the ions were independent of the current-strength, but dependent upon the concentration.

As solutions more and more dilute were examined, he found, however, that a point was finally reached beyond which further dilution caused no change in the relative rates of migration. This is not difficult to understand. In the concentrated solutions there is a considerable quantity of undissociated molecules, and these molecules offer resistance to the motion of the ions among them, which resistance must be dependent upon the nature of the ions, the molecules, etc. As the dilution becomes greater the molecules disappear, until the point is reached where the dissociation is complete, or their effect not to be observed.

Hittorf did not discover any effect produced by the moderate changes of temperature to which his solutions were subjected. Later and more extended investigations, however, have proved that the relative rates of migration of the ions are slightly affected by changes in the temperature, and that these changes lie in such a direction that with increased temperature the velocities of the different ions seem to tend toward a common value.

If other solvents than water be used, as, for instance, methyl or ethyl alcohol, in which dissociation also takes place, the values found for aqueous solutions are no longer applicable.

Thus far only univalent ions have been considered. In the case of ions of other valencies, the method for determination of the shares of transport is the same as given. If one bivalent is combined with two univalent atoms, as in $BaCl_2$, the charged ions are Ba^{II} and Cl, Cl, and, in the above notation, $\frac{n}{1-n}$ represents the ratio existing between the rates of migration of the barium ions Ba^{II} and that of the two Cl ions.

Still another advance was possible through Hittorf's study of the concentration-changes at the electrode; namely, the compositions of the ions resulting from the dissociation of the compounds were discovered. Cyanide of silver, for example, dissolves in a solution of potassium cyanide, but the exact nature of the compound formed, and the composition of the ions into which it dissociates, cannot be determined from this fact alone. When Hittorf conducted the electric current through this solution, silver was precipitated at the cathode. Upon analysis of the solution at the cathode after the electrolysis he found that the quantity of potassium had increased by an amount equivalent to the separated silver, and also equivalent to the quantity of electricity passed through the circuit as measured in a silver voltameter. This result he interpreted in the following manner: K is the positive and $Ag(CN)_2$ the negative ion. Leaving out of account the precipitated substance, the positive and negative ions must always be present in equivalent amounts, which evidently requires the silver and potassium to be present here in equivalent quantities. The potassium must have separated at the cathode, and then have acted upon the solution, precipitating silver and entering the solution itself. This explains the presence of an excess of potassium in this part of the solution exactly equivalent to the separated silver or to the electricity which has passed. The precipitation of the silver then becomes, in this case, a secondary reaction; the potassium equivalent to the electricity has precipitated the equivalent of silver.

In a similar manner he found that sodium platinic chloride dissociates into two sodium cathions and the bivalent anion $PtCl_6$, sodium gold chloride into one sodium ion and the univalent ion $AuCl_4$, ferro-cyanide of potassium into four potassium ions and the quadrivalent ion $FeCy_6$, ferri-cyanide of potassium into three potassium ions and the trivalent ion $FeCy_6$, etc.

These conclusions of Hittorf, which, when first published, met with great opposition, are now known to be perfectly correct, and have been proved in many ways, for instance, by the method of the determination of freezing point reduction. Another experiment which Hittorf also made may well be considered at this point. He found in his study of potassium chloride and iodide solutions that the chlorine and iodine ions have nearly the same velocity of migration. With our present knowledge we can safely predict that through the electrolysis of a mixture of these salts the ratio of the concentrations of chlorine and iodine will remain unchanged in all parts of the solution, since these two take part equally in the conduction. This was found to be the case. At that time this point caused some trouble, because the iodine alone separates at the electrode, and the difference between the phenomena of electrical conduction and precipitation was not understood.   Since the iodine alone separated, it was concluded that possibly this alone, belonging to the easier decomposed body, conducted the electricity.   The fact that iodine alone separates at the electrode in this case is no evidence as to what ions conduct the electricity through the solution.

F. Kohlrausch[1] has lately arranged the relative rates of migration of the ions of the most important and best - investigated electrolytes in tabular form, and this table is here given.   The values are the transference numbers, and represent the ratio of the velocity of migration of the anion to the sum of the velocities of the anion and cathion.   The concentrations of the solutions (m) are given in gram-equivalents per liter.

[1] *Wied. Ann.* lv. 287, 1893.

[TABLE

TABLE I.—Hittorf's Transference Numbers for the Anions.

$m \frac{\text{gm. equiv.}}{\text{liter}}$ =	0·01	0·03	0·05	0·1	0·2	0·3	0·5	0·7	1	1·5	2	2·5	3	4	5	6	8	10	20
KCl	—	0·503	0·509	**0·507**	0·512	0·512	—	0·514	—	—	0·516	0·514	—	—	—	—	—	—	—
Kuschel	—	0·530	—	—	—	0·505	—	0·517	—	—	0·521	—	—	—	—	—	—	—	—
NaCl	—	—	0·02	0·63	0·63	—	—	0·63	—	—	—	—	0·65	0·65	0·65	—	—	—	—
LiCl k	—	—	0·68	0·70	0·72	0·75	—	0·73	0·74	—	0·74	—	0·75	—	—	0·77	—	—	—
NH₄Cl	—	—	—	0·508	—	—	—	0·514	—	0·514	—	0·514	0·517	—	—	—	—	—	—
½BaCl₂	—	—	0·61 b	0·61	0·58 b	—	—	—	—	—	—	—	—	—	—	—	0·77	—	—
½CaCl₂	—	—	—	0·68	0·68	—	—	—	0·69	—	—	—	—	—	—	0·75	0·81	0·79	—
½MgCl₂	—	—	—	0·68	0·68	—	—	—	0·71	—	—	—	—	—	—	—	0·32	—	—
HCl	0·21	—	—	0·21	0·17	0·16	—	0·17	—	—	0·19	0·19	—	0·725	0·78	—	—	—	—
½CdJ₂	—	0·61	0·65 b	0·68	0·88	0·93	0·93	—	1·11	1·17	1·14	—	1·27	—	—	—	—	—	—
Lenz	—	—	—	—	0·80	—	—	—	1·04	—	—	—	—	—	—	—	—	—	—
KNO₃	—	0·50	—	0·50-	—	0·49	—	—	0·49	—	0·48	—	—	—	—	—	—	—	—
NaNO₃	—	—	—	0·61	0·61	0·61	0·52	—	0·50	—	0·47	—	0·60	—	0·59	—	—	—	—
AgNO₃	0·526	0·524	0·626	0·526	—	0·525	—	—	—	—	—	—	—	—	—	—	—	—	—
L. u. Nernst	0·526	0·526	0·528	0·529	—	—	—	—	—	—	—	—	—	—	—	—	—	—	—
½Ca(NO₃)₂	—	—	—	—	—	0·65	—	—	—	—	—	—	—	—	—	0·71	—	—	—
KClO₃	—	—	—	0·46	—	0·445	—	—	—	—	—	—	—	—	—	—	—	—	—
KC₂H₃O₂	—	—	—	0·32	0·33	—	—	—	—	0·42	—	—	—	0·41	0·33	—	—	—	—
NaC₂H₃O₂	—	—	—	0·44	—	—	—	—	0·64	—	—	—	—	—	—	—	—	—	—
½K₂SO₄	—	—	—	0·60	—	0·63	—	—	—	—	—	—	0·60	—	—	—	—	—	—
½Na₂SO₄	—	—	—	—	0·65	—	—	—	—	—	—	—	0·65	—	—	—	—	—	—
½Li₂SO₄ k	—	—	0·64	0·64	0·65	0·65	0·68	0·68	0·70	0·72	0·73	—	0·76	—	0·78	—	—	—	—
½ZnSO₄	—	—	—	0·21	0·21	—	—	—	0·17	—	—	—	—	—	—	—	—	—	—
½CuSO₄	—	—	—	—	—	—	—	0·68	0·69	0·71	0·73	—	0·76	—	—	—	—	—	—
Kirmis	—	—	—	—	—	0·18 w	—	0·68	0·17	—	0·19 w	—	0·18 w	0·18	—	—	—	—	—
½H₂SO₄	—	—	—	—	—	—	0·49	0·44	0·55	0·42	—	—	—	—	—	—	—	—	—
½K₂CO₃ k	0·30	—	—	0·37	0·52	0·41	—	—	—	0·42	—	—	0·41	—	—	—	0·34	0·27	0·40
½Na₂CO₃ k	—	—	—	0·48	0·58	—	—	—	0·55	—	—	—	0·58	—	—	—	—	—	—
½Li₂CO₃ k	—	—	—	0·59	—	—	—	—	—	—	—	—	—	—	—	—	—	—	—
KOH k	—	—	—	0·74	0·73	0·80	0·87	0·74	0·88	0·90	—	—	—	—	—	—	—	—	—
NaOH k	—	—	—	0·84	0·85	—	—	—	—	—	—	—	—	—	—	—	—	—	—
LiOH k	—	—	—	—	—	—	—	—	—	—	—	—	—	—	—	—	—	—	—
$m \frac{\text{gm. equiv.}}{\text{liter}}$ =	0·01	0·03	0·05	0·1	0·2	0·3	0·5	0·7	1	1·5	2	2·5	3	4	5	6	8	10	20

CHAPTER V

Specific and Molecular or Equivalent Conductivity.—We have already learned something of the nature of conductors of the first class. The resistance of such conductors is dependent upon the nature of the material, its form and the temperature.

Representing the resistance of a cylinder one meter long, and of one square millimeter cross-section, by k, the resistance of a similarly formed piece of the substance at the same temperature is $k \frac{l}{q}$, where l is the length in meters and q the area of the transverse section in square millimeters. The factor k represents the specific resistance of the substance.

The resistance of a cylinder of mercury one meter long and one square millimeter in section at 0° has been chosen as the unit of resistance, so that all resistances referred to this unit are easily comparable. In using this unit it may also be said that the value k for any substance is that number by which the resistance of a certain amount of mercury of a definite form at 0° must be multiplied to give as product the resistance of the substance itself when possessing exactly the same volume and form.

Besides the *mercury,* or the so-called *Siemens unit,* the *ohm* is also used as unit. This is the resistance of that length of a conductor in which the potential falls one volt when the current-strength is one ampère. The *specific resistance* expressed in ohms is evidently that of a cylinder of the substance in question, which is one meter long, and whose cross-section is one square millimeter in area. The Siemens unit stands to the ohm in the ratio 1 : 1·063, so that in order to express Siemens units in ohms, it is necessary to divide the former by 1·063. Conversely the product 1·063 × ohms gives us the value of the ohms in Siemens units. Recently the centimeter has come somewhat into use in the unit of specific resistance instead of the meter and millimeter. This specific resistance is then that of a cube whose edge is one centimeter long, and is evidently ten thousand times as small as the former.

The greater the resistance, the less is the conductivity, and as the conductivity increases, the resistance decreases. That is, resistance (R) and conductivity (L) are reciprocal values, or

$$R = \frac{1}{L}.$$

The word conductivity is used mainly with reference to solutions, and will be here so used. The word resistance is applied more especially to conductors of the first class. It seems at first natural to express the conductivity of solutions in reciprocal Siemens units or ohms, but this has not proved a satisfactory method of expressing these solution-conductivities. The conductivity depending in this case almost entirely upon the dissolved substances, it has been found advantageous to base the comparisons upon the contents of the solutions, the conductivities of solutions being compared which contain one gram-molecule of the electrolyte. This is called the molecular conductivity, and is commonly expressed by μ. If v be the volume in cubic centimeters in which one gram-molecule of an electrolyte is dissolved, then the molecular conductivity $\mu = v \times$ specific conductivity, and may be deduced as follows. Between two parallel electrodes exactly one centimeter apart—for instance, two opposite walls of a vessel—one liter of a solution containing one gram-molecule of the electrolyte is placed. The value of v the volume for the solution is then 1000, the cross-section of this liquid conductor is 1000 sq. cm., and its conductivity is called the molecular conductivity μ for the electrolyte. This value is evidently 1000 times as great as the specific conductivity l of a cube whose edge is one centimeter long, consequently $\mu = vl$. With a half normal solution ($v' = 2000$) two liters of the solution must be placed between the electrodes to introduce one gram-molecule, and the cross-section of the circuit has an area of 2000 sq. cm. The molecular conductivity μ' is then 2000 times as great as its specific conductivity l' or $\mu' = v'l'$.

The specific conductivities naturally change when the concentration of the solution is changed, as do also the molecular conductivities. In other words, μ the molecular conductivity is equal to the product vl or $10000 \, vl_1$, where l represents the specific conductivity of the solution in the form of a cube whose edge is one centimeter long, and l_1 is the specific conductivity of the column of liquid one meter long, whose area of cross-section is one square millimeter. If the number of gram-molecules of substance which are contained in one liter be represented by m, then

$$\mu = \frac{l \cdot 10^3}{m} = \frac{l_1 \cdot 10^7}{m}.$$

Instead of the molecular conductivity μ, the value λ is sometimes used as a basis to which conductivity may be referred; it is called the "*equivalent conductivity*."

$$\lambda = \frac{l \cdot 10^3}{n} = \frac{l_1 \cdot 10^7}{n}.$$

when n is the number of equivalents which a liter of the solution contains. For univalent compounds, as those first to be considered, the values of μ and λ are identical.

General Laws.—The first clear ideas to be obtained concerning the conductivity of electrolytes resulted from the experiments of Kohlrausch. The work of discovery was then rapidly pushed forward by Arrhenius, Ostwald, and others. It was found that the equivalent conductivities of all electrolytes increased with increasing dilution, reaching in many cases a maximum which did not alter for further dilution. In cases where this maximum value for the equivalent conductivity is reached, the following law of Kohlrausch [1] is true.

The equivalent conductivity of a binary electrolyte is equal to the sum of two values, one of which depends only upon the cathion, and the other upon the anion.

This expresses the fact that the conductivity of an electrolyte is an additive property; in other words, it is simply the sum of the conductivities of its ions.

---

[1] *Wied. Ann.* vi. 1, 1879, and xxvi. 213, 1885.

This law is evident from the accompanying table containing the equivalent conductivities of the compounds of the metals in the vertical columns, with the radicles in the horizontal.[1]

	K	Na	Li	NH$_4$	H	Ag
Cl	123	103	95	122	353	...
NO$_3$	118	98	...	...	350	109
OH	222	201	...	...	...	...
ClO$_3$	115	...	...	...	...	103
C$_2$H$_3$O$_2$	94	73	...	...	...	83

The differences between corresponding values of the vertical columns are approximately equal, as are also those between corresponding values in horizontal lines. This could only be explained by considering the conductivities as the sums of two independent constants. A great many other properties of dilute solutions of electrolytes, which may be similarly expressed as the sums of the properties due to the ions constituting the electrolyte, are recognised. Ostwald has called them additive properties; among these may be mentioned the colour and the index of refraction.

It is found that the dissociation theory offers a perfect explanation of the above law of conductivity. The conductivity of electricity through a solution consists in a motion of the single ions. If in a solution of x ions in an electric circuit there are 100 ions passing the cross-section of the liquid conductor in unit time, then 200 ions would pass if there were

[1] $t = 18°$ C. The numbers represent the equivalent conductivity at extreme dilution, and are taken from Kohlrausch's tables (*Wied. Ann.* 1. 406, 1893).

2 x ions in the solution, providing other conditions remained unaltered, that is, the conductivity would be twice as great as before. The conductivity of a solution depends primarily upon the number of ions between the electrodes, but also upon the sum of the velocities of migration of the two kinds of ions.

The equivalent conductivity of an electrolyte can be measured directly by placing a solution of one gram-equivalent of the electrolyte in question into a vessel, two of whose opposite walls one centimeter apart, serve as electrodes. Other dimensions of the vessel than the distance between these wall-electrodes need not be fixed. The measured conductivity is in such a case also the equivalent conductivity. The volume of the solution may evidently be of any desired magnitude, so long as the quantity of the dissolved substance is the same. If the substance is completely dissociated there are in such a case two gram-equivalents of ions between the electrodes, and the conductivity will always be the same, no matter how much of the solvent be present, while these ions which alone carry the electricity are present between the electrodes. The size of the electrodes need not be taken into consideration, for it plays no part in the conductivity, provided that the number of ions between the electrodes is not changed. Consequently with increasing dilution, a maximum for equivalent conductivity is finally obtained which remains constant for further dilution. It is also easy to understand why the equivalent conductivity should be less for concentrated solutions than for dilute, because in the former the dissociation is less; or, in other words, the number of ions has been diminished. With increasing dilution the degree of dissociation, and consequently the equivalent conductivity, also increases, until that maximum value is reached which corresponds to complete dissociation.

Through these facts we observe the superiority of the new theory over that of Clausius. According to the latter, the conductivity depends upon the frequency of the changes between the parts of molecules, and it would seem necessary to conclude that the more concentrated the solution, the more often would these changes take place ; the equivalent conductivity, in consequence, would increase with increasing concentration, which is contrary to fact.

Dilute equivalent solutions of neutral salts, strong acids, and bases, since they are practically completely dissociated, contain the same number of ions ; consequently their equivalent conductivities stand in the same ratio as the sums of the velocities of migration of their respective ions. The velocity of migration of an ion is a constant, because the ion moves independently of other ions in the solution ; therefore $\lambda = K(u + v)$, K being a factor of proportion dependent upon the chosen units, and u and v representing the rates of migration of the gram-equivalents of the positive and negative ions. This is an expression for the law of Kohlrausch.

The maximum value of the equivalent conductivity of an electrolyte is the sum of the velocities of migration of its ions, and as we have already obtained the relative values of these rates of migration from Hittorf's work (see p. 70), we can now calculate the single values.

$$K(u + v) = \lambda$$

$$\frac{u}{v} = \frac{n}{1 - n}$$

$$\overline{v \cdot K = (1 - n)\lambda}$$

$$u \cdot K = n\lambda.$$

The proportionality factor K being unity, in other words, expressing the rates of migration in the same units as the conductivities, then

$$v = (1 - n)\lambda \; ; \; u = n\lambda.$$

When the value of the rate of migration for a single ion is known, that of the others may be obtained from the shares of transport as well as from values for the limits of conductivity; the fact that the results as obtained by the two methods agree, justifies the interpretation of · the phenomena. Kohlrausch has calculated and compared many of these velocities of migration, and found that the two methods of determination give the same results.

From the latest and most correct collection of velocities of migration computed by Kohlrausch (*Wied. Ann.* 1. 385, 1893) the following mean values have been selected ($t = 18°$ C.):

K = 60, Na = 40, Li = 33, NH_4 = 60, H = 290, Ag = 52 ;
Cl = 62,   I = 63, NO_3 = 58, ClO_3 = 52, ClO_4 = 54,
$C_2H_3O_2$ = 31, OH = 165.

The conductivity at great dilution, λ, is $= u + v$. If not all the molecules were dissociated, but only half of them, the conductivity λ' would only be half as great, because it is proportional to the number of ions, or

$$\lambda' = \frac{\lambda}{2} = \frac{1}{2}(u + v),$$

and in general $\lambda_v = x(u + v)$, where λ_v represents the equivalent conductivity of an electrolyte, one gram-equivalent of which is contained in the volume v, and x is the portion which is dissociated, or the degree of dissociation. Representing the maximum value for the equivalent conductivity by λ_∞ the following equations may be written:

$$\lambda_\infty = u + v$$
$$\lambda_v = x(u + v)$$

$$x = \frac{\lambda_v}{\lambda_\infty}.$$

The degree of dissociation of a substance in a solution is equal to the ratio of the equivalent conductivity of that solution to its equivalent conductivity at infinite dilution.

As already seen (p. 59) Arrhenius had come to this conclusion, and there was agreement between the values for the degree of dissociation as obtained from the conductivity and from the method of freezing point depression.

The determination of the degree of dissociation of different substances has led to very important results. Ostwald found by experiment that the order in which the acids act in the catalysis of methyl acetate, sugar, etc., is also the order of their "affinities" for a base. This latter could be determined by thermochemical or volume-chemical measurements. From Ostwald's work a measure for the chemical activity, "affinity," or "strength" of an acid (or base) is obtained.

Arrhenius sought to discover the existence of a connection between conductivity and the chemical activity, determined as just mentioned, and found that the two are in reality closely connected. Having solutions of two acids, each containing one gram-equivalent per liter, the "strengths" will evidently not be the same if the degrees of. dissociation differ. On diluting the two solutions the dissociation increases, and finally the acids are wholly dissociated. At this point the two "strengths" must be the same. The relative "strengths" of acids and bases change therefore with the conditions of concentration. This was shown by Ostwald before the advent of the Arrhenius theory of dissociation.

Determination of the Dissociation Constant by Electrical Conductivity — Kohlrausch Method. — Recognising the dissociation theory, and the applicability of the gas laws to dissolved substances, as established by van't Hoff, an affinity constant for binary electrolytes, or, in other words, a dissociation constant which is independent of the dilution, may be calculated. This was first shown by Ostwald.[1]

According to the law concerning the effect of mass in a reaction for a gas which decomposes into two parts, the temperature being constant, the following principle holds :—

The product of the concentrations of the two parts, divided by the concentration of the undissociated part, is a constant.

For example, the vapour of ammonium chloride at high temperature decomposes partly into ammonia and hydrochloric acid. When the temperature is constant the conditions may be represented by $\frac{p_1^2}{p} = K'$, where p_1 is the pressure due to the ammonia (or acid), the two being the same, because the substances are present in equal molecular quantities. p represents the pressure due to the ammonium chloride. These values, being evidently proportional to the concentrations, may be used in place of them.

Compressing the gas, the values for the partial pressures are increased, and increasing its volume, the partial pressures diminish, but the value K' remains

[1] *Zeitschr. physik. Chem.* ii. 270, 1888.

constant. Moreover, if an excess of one of the components is added, no change takes place in the constant. If ammonia is added, its partial pressure or the ammonia pressure is increased, and if no other change took place, the value of K' would be greater than before; but, since this remains the same, either the numerator of $\dfrac{p_1^2}{p}$ must decrease or the denominator increase, or both changes take place. The latter happens. A part of the ammonia and acid combine to form ammonium chloride, this combination taking place to such an extent, that the product of the partial pressures of NH_3 and HCl, divided by the partial pressure of the salt, assumes its previous value :

$$\frac{p'_{NH_3} \cdot p''_{HCl}}{p'''_{NH_4Cl}} = K'.$$

Here, of course, p' and p'' do not have the same value. Since, according to van't Hoff's theory, substances in dilute solution obey the laws governing gases, it is natural to assume that such relations as are illustrated by the ammonium chloride vapour would be found in the case of a binary electrolyte, or one which decomposes in solution into two ions. For example, a dilute aqueous solution of acetic acid contains besides undissociated CH_3COOH also a quantity of the two ions H and CH_3COO, and it should consequently be expected that the equation $\dfrac{c_1^2}{c} = K$ for this solution would be independent of the dilution, c_1 representing the concentration of each of the ions, and c that of the undissociated molecule. Such is actually the case. The presence of other substances in the solution does not affect this constant. The magnitude of the dissociation constant K is a characteristic of every compound, and its determination is therefore of great importance.

In order to show the existence of this relation for solutions, it is of course primarily necessary to have a method for finding out the concentrations c and c_1. For this purpose the determination of electrical conductivity is most satisfactory, and it is in consequence of this fact that the conductivity measurements in general are of such value.

If in V liters one gram-molecule of a binary electrolyte is dissolved, and x is the degree of dissociation of the electrolyte, or that part of the gram-molecule which has decomposed into ions, $\frac{x}{V}$ will then stand for the quantity of a gram-ion in one liter, that is, for the concentration of each of the ions; therefore $\frac{1-x}{V}$ must be the concentration of undissociated material, and finally $\frac{x^2}{(1-x)V} = K$.

Hence for the determination of the dissociation constant K of a given solution, only the degree of dissociation need be known. But this dissociation x is equal to the ratio of the molecular or equivalent conductivity of the solution in question to the same at infinite dilution, or $x = \frac{\lambda_v}{\lambda_\infty}$, and substituting this value of x in the above equation,

$$\frac{(\lambda_v)^2}{(\lambda_\infty)^2\left(1 - \frac{\lambda_v}{\lambda_\infty}\right)V} = \frac{(\lambda_v)^2}{\lambda_\infty(\lambda_\infty - \lambda_v)V} = K.$$

A knowledge of the molecular or equivalent conductivity and the conductivity at infinite dilution suffice therefore for a determination of the dissociation constant K.

Before considering the experimental proof of the formula it is advisable to become acquainted with the methods for determining the conductivity of a solution.

Through the use of Ohm's law $\left(C = \frac{\pi}{R}\right)$ measurement of the resistance of metallic conductors, or those of the first class, is very simple, but in the case of electrolytes this is no longer true. The gradual fall of potential, π, which exists in that portion of the circuit occupied by a solution is more difficultly determinable, because of the variable changes of the potential at the electrodes due to the chemical processes taking place there. Many methods have been devised for overcoming this difficulty and measuring the actual conductivity or resistance of solutions. The one now commonly in use is the only one which need be here described. This is the method of Kohlrausch.

The method depends upon the application of an alternating current. By its use the change of potential at the electrode due to polarisation is practically removed, for the polarisation effect produced by the current of one direction for a small fraction of a second, may be considered as neutralised by the current when its direction is reversed. The circuit is then similar to one composed of conductors of the first class only, and has practically only constant electromotive forces at the electrodes. Under such conditions, the resistance may be measured as in the case of metallic conductors.

The apparatus employed is shown in Fig. 16, and

is essentially the ordinary Wheatstone's bridge arrangement.[1] At 5 is a galvanic element connected with an

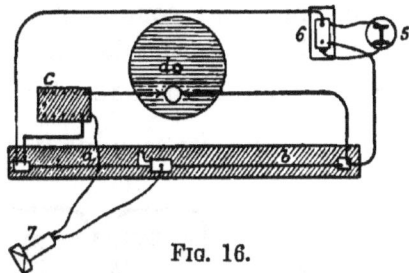

induction coil (6) for
producing the alter-
nating current. At 7
is a telephone, a gal-
vanometer being, of
course, inapplicable to
alternating currents.
The four arms of the
bridge are represented

FIG. 16.

by a, b, c, and d. When the element at 5 is in
activity a tone is audible in the telephone, except
when the resistances of the four parts of the bridge
are in the relation expressed by $\frac{a}{b} = \frac{c}{d}$. When
such is the case, a clearly defined minimum, if not
entire absence of sound, is the result. At c there
is a resistance box usually measuring ohms or
Siemens units. The electrolyte whose resistance is
to be determined is placed in the vessel at d, which
is in a thermostat. The two parts of the bridge
a and b are most conveniently composed of a
uniform platinum wire, which is stretched over a
meter scale; upon this wire a movable contact connected with the telephone is brought by changing the
position of which the tone-minimum may be found.
Resistance of such an amount is introduced into the
branch c of the circuit, that the point of contact for
minimum tone is in the middle portion of the wire a,
b. That portion of the wire from one end to the point
of contact leading to the telephone is a, and the other
portion b, and the lengths of these may be read to

[1] Ostwald, *Zeitschr. physik. Chem.* ii. 561, 1888.

tenth-millimeters directly from the scale. The actual resistance of this wire does not enter into the calculation, as it is only the ratio $a : b$ which is required. The resistance of the other metallic conductors in the circuit from the resistance box to the electrolyte, etc., must be negligibly small. The desired resistance of the electrolyte is then $d = \dfrac{cb}{a}$ and its conductivity

$$\frac{1}{d} = L = \frac{a}{cb}.$$

For a vessel in which to determine the conductivity the one represented in Fig. 17 is usually employed.

The electrodes may be separated by the desired distances, and their area made to suit the requirements of most cases. Platinised platinum electrodes are used to the best advantage.

Expressing the distance between the two electrodes in meters and their surface-area in square millimeters, the specific conductivity may be obtained as follows:

FIG. 17.

$$\frac{1}{d} = L = \frac{a}{cb} = l_1 \frac{\text{sq.mm.}}{\text{meters}},$$

and also, according to page 82, the molecular or equivalent conductivity $\lambda = \dfrac{l_1 \cdot 10^7}{n}$. The conductivity is thus expressed either in reciprocal ohms (mhos) or in Siemens' units.

In order to obviate the necessity of measuring the space between the electrodes, mercury may be placed in the vessel and its resistance measured at 0°, expressed in any desired units. Now measuring

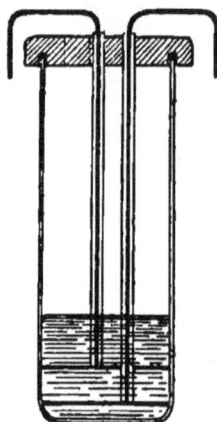

the resistance of an electrolyte in the same vessel, the two resistances found will stand in the same ratio as the specific resistances of the mercury and electrolyte. The specific resistance of mercury (referred to one meter length with one square millimeter sectional area) is one Siemens unit; therefore the relation of the desired resistance to that of the mercury (or the specific resistance of the electrolyte) is expressed in Siemens units. The resistance of the mercury in the vessel at 0° is called the resistance-capacity of the vessel, and, as stated, may be measured in any desired units, as the final result is only a ratio. The units chosen must, however, be retained in measuring the · resistance of the electrolyte. The resistance-capacity is then represented by that number by which the resistances measured in the vessel must be divided in order to obtain the specific resistance, or it is the number with which the conductivity corresponding to the measured resistance must be multiplied in order to obtain the specific conductivity.

In practice this method is not always applicable, because the resistance offered by the mercury between the electrodes is so small that it cannot be measured with sufficient accuracy. This is true in the case of the vessel above depicted, which is in most common use. To obtain the resistance-capacity of such a vessel, a solution possessing much greater specific resistance than mercury is used, its specific resistance being measured in turn, in a vessel of different form. The solution best adapted for such a purpose is 0.02 normal potassium chloride. The relation which its specific conductivity bears to the conductivity as measured in the vessel, is the resistance-capacity of the vessel.

The calculation is perhaps simplified when carried out in the manner usually adopted, as follows.   If c be the resistance in the resistance box, and a and b the lengths of the portions of the platinum wire as determined by the movable contact when the minimum tone is given by the telephone, the conductivity of the electrolyte is then expressed by

$$L = \frac{a}{bc}.$$

Knowing the resistance-capacity of the vessel to be κ, the specific conductivity l_1 of the electrolyte would be, as previously explained,

$$l_1 = \kappa \frac{a}{bc} \text{ reciprocal Siemens units.}$$

The equivalent conductivity would therefore be represented by

$$\lambda = \kappa \frac{a}{b \cdot c \cdot n} 10^7.$$

Instead of using $\frac{1}{n}$ where n is the number of gram-equivalents in a liter of solution, V may also be used where V is the number of liters in which one gram-molecule is dissolved, and using ξ instead of $\kappa \cdot 10^7$, then

$$\lambda = \xi \frac{a \cdot V}{b \cdot c} \text{ reciprocal Siemens units.}$$

The value of ξ may be calculated, the equivalent conductivity λ of a 0·02 normal potassium chloride solution being known, and the value $\frac{a}{b \cdot c}$ may be experimentally determined.   V is here 50, and ξ, being the only unknown quantity in the equation, is determined. This ξ is equal to the resistance κ of the vessel multiplied by 10^7.

Having once determined the value of ξ for the vessel, one can proceed to measure the resistance of any electrolyte at any known state of dilution, and calculate its equivalent conductivity by means of the above formula.

The specific conductivity of a 0·02 normal potassium chloride solution (referred to a length of one meter with one square millimeter section) is $l_1 = 2 \cdot 244 \cdot 10^{-7}$ reciprocal Siemens units at 18°, or $2 \cdot 594 \cdot 10^{-7}$ at 25°. The corresponding equivalent conductivities are:

$$\lambda \begin{cases} = 2 \cdot 244 \cdot 10^{-7} \cdot 50 \cdot 10^7 = 112 \cdot 2 \text{ at } 18°, \\ = 2 \cdot 594 \cdot 10^{-7} \cdot 50 \cdot 10^7 = 129 \cdot 7 \text{ at } 25°. \end{cases}$$

The other specific conductivity l is $= 2 \cdot 244 \cdot 10^{-3}$ and $2 \cdot 594 \cdot 10^{-3}$ at these respective temperatures from which naturally the same values for λ result as above:

$$\lambda \begin{cases} = 2 \cdot 244 \cdot 10^{-3} \cdot 50 \cdot 10^3 = 112 \cdot 2 \text{ at } 18°, \\ = 2 \cdot 594 \cdot 10^{-3} \cdot 50 \cdot 10^3 = 129 \cdot 7 \text{ at } 25°. \end{cases}$$

That the equivalent conductivities are very great is shown by the above. A gram - equivalent of potassium chloride in the 0·02 normal solution, when placed between electrodes one centimeter apart, offers at 25° a resistance to the electric current of only $\dfrac{1}{129 \cdot 7}$ Siemens units.

The equivalent conductivities of all binary electrolytes for infinite dilution are of about this order, and vary between 50 and 500. This conductivity λ may evidently have exceedingly small values when the concentration of the solution is great.

Besides the value of λ_v it is necessary to know the magnitude of λ_∞, in order to calculate the dissociation constant K of an electrolyte.

$$K = \frac{(\lambda_v)^2}{\lambda_\infty(\lambda_\infty - \lambda_v)V}.$$

Having learned how λ_v is determined, attention will now be given to λ_∞. In some cases this is obtained in the course of determining λ_v in dilute solutions, a maximum value for λ_v being obtained, beyond which further dilution does not affect its value. This maximum is then λ_∞. This is a method applicable only to electrolytes whose tendency to dissociate is great. The direct experimental determination of λ_∞ is not possible for compounds which dissociate but little, because a condition of complete decomposition would only be reached at such a great dilution that the measurement of the resistance would be impossible. It is therefore inapplicable, for instance, to solutions of the organic acids and bases, where the knowledge of the value of λ_∞ is very important. Fortunately, however, the alkali salts of all acids and the halogen-acid salts of all bases, whether strong or weak, dissociate to a high degree in moderately dilute solutions, so that the value of λ_∞ can be determined for these compounds. But λ_∞ represents the sum of the velocities of migration of the ions of the electrolyte, and since the rate of migration of the alkali metal ion in one case, and of the halogen ion in the other, is known, that of the acid and basic ion respectively may be obtained by subtraction of these quantities from the values of λ_∞. Having thus obtained the velocities of the negative ion of the metal compound, and of the positive ion belonging to the halogen compound, it is only necessary to add to the former the rate of migration of the hydrogen ion, and to the latter that of the hydroxyl ion, both of which are easily determined experimentally, and the values of λ_∞ for the desired compounds are determined.

Through the investigation of a great many feebly dissociating acids and bases, under very varying conditions of dilution, it has been found that there exists for each a dissociation constant which is independent of the dilution, in conformity with the theoretical predictions alluded to above.

As a consideration of the significance of this constant belongs to the subject of chemical statics, it will not be discussed in detail here, but it may well be mentioned that the order of the magnitudes of these constants for different compounds is also the order of their degrees of dissociation, as they exist in solutions of the same equivalent concentration. Direct proportionality does not, however, exist between the constants and the dissociation, for as the dilution is made greater, the degrees of dissociation approach a common value. Some of the results of the existence of these constants as they were empirically established by Ostwald before the dissociation theory was proposed will now be considered.

1. With increasing magnitude of V in the formula

$$\frac{(\lambda_v)^2}{\lambda_\infty(\lambda_\infty - \lambda_v)} = VK$$

the value of the left-hand portion of the equation must also increase and approach infinity. As λ_v and λ_∞ are always finite quantities, this can only occur when $\lambda_v = \lambda_\infty$ or *with increasing dilution the equivalent conductivity approaches the value* λ_∞.

2. In the case of weakly dissociating and, consequently, badly conducting electrolytes, where λ_v is very small as compared to λ_∞, the value of $\lambda_\infty - \lambda_v$ is only slightly changed by increasing dilution, so that it may be practically considered as constant. As a result

$$\frac{(\lambda_v)^2}{V} = \text{constant.}$$

That is, the equivalent conductivity increases with increasing dilution in proportion to the square root of the volume, or the square of the equivalent conductivity increases in proportion to the volume.

3. If the formula for the dissociation be written as follows :

$$\frac{x^2}{(1-x)V} = K,$$

the value of $1 - x$ for slightly dissociating substances differs but little from 1, and the equation approaches the form

$$\frac{x^2}{V} = K.$$

4. When the volume V of two or more slightly dissociating electrolytes is the same, then from the above

$$\frac{(\lambda_v')^2}{(\lambda_v'')^2} = \text{constant, and}$$

$$\frac{x_1^2}{x_2^2} = \frac{K'}{K''},$$

λ_v', x_1, K', and λ_v'', x_2, K'' being the molecular conductivities, degrees of dissociation, and dissociation constants. $\lambda_v = x \cdot \lambda_\infty$, and in consequence, when the maxima of conductivity of two electrolytes are practically equal, which is the case with many acids because of the very great velocity of migration of the hydrogen ions common to them, the equation

$$\frac{(\lambda_v')^2}{(\lambda_v'')^2} = \frac{K'}{K''} \text{ is true, or}$$

the squares of the equivalent conductivities of different electrolytes stand to one another in the same proportion as the dissociation constants, when the degrees of dilution are the same.

5. In the formula

$$\frac{(\lambda_v)^2}{\lambda_\infty(\lambda_\infty - \lambda_v)V} = K,$$

λ_v, in the case of highly dissociating electrolytes, may be considered as nearly constant with changing dilution, and as λ_∞ is *eo ipso* constant,

$$\frac{1}{(\lambda_\infty - \lambda_v)V} = \text{constant.}$$

The product of the difference between the maximum of conductivity and the equivalent conductivity into the volume is a constant.

6. The equation

$$\frac{x^2}{(1-x)V} = K$$

may be considered as having the form

$$\frac{1}{(1-x)V} = K$$

for highly dissociating electrolytes, for here x^2 is nearly equal to 1.

The undissociated portion of the compound, multiplied by the volume, is equal to the reciprocal value of the dissociation constant.

From this it is evident that, if the undissociated portion amounts to 1% when $V = 500$, it would decrease to $\frac{1}{2}\%$ for $V = 1000$.

7. If two or more electrolytes having great tendencies to dissociate are compared under the same condition of dilution,

$$\frac{\lambda_\infty' - \lambda_v'}{\lambda_\infty'' - \lambda_v''} = \text{constant, and}$$

$$\frac{(1 - x')}{(1 - x'')} = \frac{K''}{K'};$$

the latter may be verbally expressed as follows:

The undissociated portions of different electrolytes, at the same degree of dilution, are inversely proportional to the dissociation constants.

If the different maxima of conductivity are nearly equal,

$$\frac{\lambda_\infty' - \lambda_v'}{\lambda_\infty'' - \lambda_v''} = \frac{K''}{K'}$$

is approximately correct, or:

The differences between the maxima of conductivity and the equivalent conductivities are inversely proportional to the corresponding dissociation constants when the degrees of dilution are the same.

8. Finally, the following regularities for all electrolytes may also be deduced. With two electrolytes of the same degree of dissociation, the left side of the expression $\dfrac{x^2}{1 - x} = V \cdot \kappa$ being the same for both, the right side must also be the same, or $V' \cdot \kappa' = V'' \cdot \kappa''$, or $\dfrac{V'}{V''} = \dfrac{\kappa''}{\kappa'}$.

The dilutions at which different electrolytes possess the same degree of dissociation (and also often nearly the same equivalent conductivity) are in a constant ratio, and this is, in fact, the inverse ratio of the dissociation constants.

The foregoing approximations may often be used with advantage.

Relation between Dissociation Constants and Chemical Constitution. — Some very interesting relations have been brought to light between the magnitudes of the dissociation constants and the chemical constitution of the acids, as may be illustrated by a few examples. The constants for acetic ·and the three chloracetic acids at 25° are as follows :

Acetic acid	0·00180
Monochloracetic acid . .	0·155
Dichloracetic acid . . .	5·14
Trichloracetic acid . . .	121·

Through displacement of hydrogen by chlorine, a large increase in the value of the constant takes place, the first chlorine atom increasing the constant about 86 times. The second chlorine makes this new constant about 33·2 times greater, and the third 23·5 greater still. We are forced then to conclude that the introduction of chlorine, for example, into acetic and monochloracetic acid does not produce like effects. This is not surprising since a chlorine atom is already present in the latter compound.

An increase in the value of the constant indicates a greater degree of dissociation for the new compound, that is, its acid character is more marked. Therefore an influence in this direction must be ascribed to the introduction of chlorine. The introduction of such negative radicals as Br, Cy, SCy, OH, etc., also increases the so-called acid properties like chlorine.

The a and β substituted derivatives of acids possess very different dissociation constants, and the constitutive property of this constant is thus very marked. The same thing applies to the isomeric benzol derivatives, for example:

Benzoic acid, C_6H_5COOH	. .	0·0060
o-Oxybenzoic acid, $C_6H_4(OH)COOH$		·102
m-Oxybenzoic acid, $C_6H_4(OH)COOH$		·0087
p-Oxybenzoic acid, $C_6H_4(OH)COOH$		·00286.

These examples show that a knowledge of the dissociation constants is of aid in determining the chemical constitution of compounds. By the introduction of OH into benzoic acid ortho to the carboxyl group, the value of the constant for this acid is increased seventeen fold. When the OH enters into the meta position instead of ortho, the change from the benzoic acid value is very small but positive, while an entrance into the para position causes a considerable reduction of the constant. Consequently it might be assumed that on starting with ortho-oxybenzoic acid and substituting its different replaceable hydrogen atoms by hydroxyl, the values for the dissociation constants of the resulting compounds, though not the same as in the benzoic acids, would still show similar relations. This is the case, as may be observed from the following table:

Ortho-oxybenzoic (salicylic) acid .	.	0·102
Oxysalicylic acid, $C_6H_3(OH)_2COOH$	(2·3)	0·114
,, ,, ,,	(2·5)	0·108
a-Resorcylic ,, ,,	(2·4)	0·052
β- ,, ,, ,,	(2·6)	5·0

In the acid 2·3 as also in the acid 2·5 the new OH is in the meta position to the carboxyl group, and consequently only a very slight increase in the dissociation constant is to be expected. This agrees with the experimental observation.

In the compound 2·4 the new OH has the para position, and again a new constant less than the original one is the result. Finally, when the second OH has taken the other ortho position, as in the acid 2·6, a correspondingly great increase in the constant results, the value being about fifty times as great as before.

Velocity of Migration of Single Ions.—Conductivity measurements have served not only to determine the dissociation constants of a great many organic acids, but have also given us the relative rates of migration of the organic anions and cathions. It has already been stated that the alkali salts of the acids and the chlorides or nitrates of the bases are so highly dissociated in solution that the value of λ_∞ is experimentally determinable. By subtraction of the known velocity of migration of the metal ion or of the halogen (or NO_3) ion, we obtain for remainder the velocity of the other ion of the compound, as already explained on page 97.

Through the stochiometric comparison of the numbers representing the migration velocities of the individual ions, certain relations have been discovered, of which a few at least will be mentioned here. These are taken from Bredig's [1] work.

The velocity of migration of the elementary ions is a function of the atomic weight, and in each series of related elements the velocity increases with it. With respect to these the rule holds that great differences occur only with the first two or three members of

[1] *Zeitschr. physik. Chem.* xiii. 191, 1894.

each series. Similar or related elements with atomic weights above 35 have about the same velocity of migration. These points may be illustrated by the following data (for 25°):

Fl 50·8	Li 39·8
Cl 70·2	Na 49·2
Br 73·0	K 70·6
J 72·0	Cs 73·6

For the complex ions the following principles have been established:

Isomeric ions migrate at the same rate, *e.g.*

Butyric acid ion	30·7	Propylammonium ion	40·1
Isobutyric acid ion	30·9	Isopropylammonium ion	40·0
Cinnamic acid ion	27·3	Chinolinmethylium ion	36·5
Atropic acid ion	27·1	Isochinolinmethylium ion	36·6

Similar changes in the composition of analogous ions produce alterations (*da*) in their velocities of migration (*a*) which are of the same general order and sign, but their magnitudes are less the lower the migration rates themselves, instead of remaining constant. In other words, the velocities of migration of very complicated ions tend towards a common limiting value when the number of atoms increases. This minimum rate of motion for the univalent anions and cathions lies between 17 and 20 reciprocal Siemens units, *e.g.*:

	a	*da* for $+CH_2$
Ammonium ion, NH_4	70·4	
Dimethylammonium ion, C_2H_8N	50·1	$-2 \times 10·2$
Diethylammonium ion, $C_4H_{12}N$	36·1	$-2 \times 7·0$
Dipropylammonium ion, $C_6H_{16}N$	30·4	$-2 \times 2·9$
Dibutylammonium ion, $C_8H_{20}N$	26·9	$-2 \times 1·8$
Diisoamylammonium ion, $C_{10}H_{24}N$	24·2	$-2 \times 1·4$

In analogous series of anions and cathions of the same valency, the rates of migration are diminished:

By the addition of hydrogen, carbon, nitrogen, chlorine, and bromine.

By the displacement of hydrogen by chlorine, bromine, iodine, etc.

In general, the more complicated the ion, the lower is its velocity of migration, and in accordance therewith the polymeric ion moves slower than the simple one. This additive nature of the velocities is often obscured by considerable constitutive influences :

Metameric ions, for example, very often possess different rates of migration because of their different structures, the migration velocity increasing with increasing symmetry, *e.g.* the velocity of migration increases in passing from the primary form to the secondary, the secondary to the tertiary, etc., as seen in the following table :

Primary base, Xylidine ion, $C_8H_{12}N = 30 \cdot 0$
Secondary base, Ethylaniline ion, $C_8H_{12}N = 30 \cdot 5$
Tertiary bases $\begin{cases} \text{Dimethylaniline ion, } C_8H_{12}N = 33 \cdot 8 \\ \text{Collidine ion, } C_8H_{12}N = 34 \cdot 8 \end{cases}$
Quaternary bases $\begin{cases} \text{Picolineethylium ion, } C_8H_{12}N = 35 \cdot 1 \\ \text{Lutidinemethylium ion, } C_8H_{12}N = 35 \cdot 2 \end{cases}$

Thus the additivity, particularly with cathions, is often destroyed by the opposing influences of such constitutional differences. Indeed, the sense of the additive change may be reversed through over-compensation, *e.g.* :

Triethylammonium ion, NC_6H_{16} . $32 \cdot 6$
Methyl-triethylammonium ion, NC_7H_{18} $34 \cdot 4$

In spite of the increase CH_2 no retardation takes place, but, on the contrary, an acceleration.

Empirical Rules.—It is evident, from a considera-

tion of its derivation, that the formula $\dfrac{\lambda_v^2}{\lambda_\infty(\lambda_\infty - \lambda_v)V} = K$
is only applicable in the case of binary electrolytes.
From the fact that other acids, the dibasic, tribasic,
etc., show agreement with this formula until they are
about 50 per cent dissociated, we conclude that at
first the dissociation taking place is simply the separation of one hydrogen atom as ion from the molecule,
the rest forming the negative ion.   On continued
dilution further production of hydrogen ions takes
place, and at the same time the valency of the negative
radical increases.   Experiments have not been made
for determining a dissociation constant for ternary
electrolytes; moreover, as will be seen from the following, these would not have much value.

The above dissociation formula does not represent
the exact truth for highly-dissociated binary electrolytes, as neutral salts, mineral acids, and inorganic
bases.   An explanation of this fact is not certainly
known.   Noyes and Abbot[1] have lately shown that the
law of mass - action, as applied to the dissociation
especially of neutral salts, cannot be considered
perfectly valid; consequently we must only accept
the relations formerly deduced from the formula as
a first approximation.   On the other hand, Ostwald[2]
discovered an empirical rule governing the changes of
the equivalent conductivity of neutral salts with the
dilution, and by its aid we may calculate the basicity
of an acid as well as the value of its limiting conductivity λ_∞.   This is of great value for such salts as only
undergo complete (dissociation at very great dilutions.
It has been found that the equivalent conductivities

---

[1] *Zeitschr. physik. Chem.* xvi. 125, 1895.
[2] *Ibid.* i. 109, 529, 1887 ; ii. 901, 1888.

of the sodium salts of all monobasic acids increase by about 10 units between the concentrations $V = 32$ and $V = 1024$, while for dibasic acids these values increase by about 20, and for tribasic by 30 units. Representing this increase by Δ, and by n the basicity of the acid, $n = \dfrac{\Delta}{10}$. The following values have been obtained for Δ :

			Δ
Sodium salt of		nicotic acid .	$10\cdot4 = 1 \times 10\cdot4$
,,	,,	chinoline acid . .	$19\cdot8 = 2 \times 9\cdot9$
,,	,,	pyridine tricarbonic acid .	$31\cdot0 = 3 \times 10\cdot3$
,,	,,	pyridine tetracarbonic acid .	$40\cdot4 = 4 \times 10\cdot1$
,,	,,	pyridine pentacarbonic acid .	$50\cdot1 = 5 \times 10\cdot0$

For strongly dissociating neutral salts generally, the following relations have been shown to exist, when λ_v is not very different from λ_∞ :

$$\lambda_\infty - \lambda_v = n_1 \times n_2 \times C_v$$
$$\lambda_\infty = n_1 \times n_2 \times C_v + \lambda_v.$$

n_1 and n_2 represent the valency of the anion and cathion, while C is a constant common to all electrolytes and dependent upon the dilution. Having determined the value of C at different dilutions once for all for a single electrolyte, whose λ_∞ is known, we are able to calculate λ_∞ for other electrolytes from a knowledge of n_1, n_2, and the equivalent conductivity for a concentration at which their C is also known. If we say that $n_1 \times n_2 \times C_v = d_v$, then

$$\lambda_\infty = d_v + \lambda_v.$$

The following table of Bredig (*l.c.*) contains values of d_v for the valency products and dilutions at 25°, which come into consideration :

Valency $n_1 \cdot n_2$	d_{64}	d_{128}	d_{256}	d_{512}	d_{1024}
1	11	8	6	4	3
2	21	16	12	8	6
3	30	23	17	12	8
4	42	31	23	16	10
5	53	39	29	21	13
6	(60)	48	36	25	16

It is well to note finally that in the calculation of the value of λ_∞ for complex anions or cathions, we can use the previously indicated fact that their rates of migration depend chiefly upon the number of atoms contained in the complex ions. If it is known, for example, that the anion of a certain acid contains 18 atoms, its value of λ_∞ may be considered to be the same as that of another anion of 18 atoms without introducing any considerable error.

Conductivity and Degree of Dissociation of Water.—Thus far it has been assumed that the observed conductivity of an aqueous solution was entirely due to the material dissolved, and that the water itself possessed no conductivity. Strictly taken, this assumption is not true, for the water dissociates, though to an extremely slight degree, into H and OH ions, which take part in the conductivity. For ordinary measurements in the conductivity of solutions, the conductivity of the water is quite inappreciable. But the presence of impurities in the water, such as traces of salts, acids, and bases, which are extremely difficult to remove, may cause a considerable error in the determinations, especially when the conductivity of very dilute solutions is being determined. In such cases it is necessary to determine the conductivity of the water used and to apply a correction.

Kohlrausch has expended a great deal of effort during the past few years in determining the actual conductivity of perfectly pure water. For water which was prepared with the greatest care, he found the values of the specific conductivity in reciprocal Siemens units to be:[1]

$$l_1 \begin{cases} = 0\cdot014 \times 10^{-10} \text{ at } \quad 0° \\ = 0\cdot040 \quad ,, \quad ,, \quad 18° \\ = 0\cdot058 \quad ,, \quad ,, \quad 25° \\ = 0\cdot089 \quad ,, \quad ,, \quad 34° \\ = 0\cdot176 \quad ,, \quad ,, \quad 50° \end{cases}$$

" One millimeter of this water at 0° had a resistance equivalent to that of forty million kilometers of copper wire of the same sectional area—an amount of wire capable of encircling the earth a thousand times."

For reasons not necessary to give here, it is probable that this experimentally-found value of Kohlrausch represents the actual conductivity of pure water. On this basis we can easily determine the degree of dissociation of the water. As the above table indicates, the conductivity of a column of water one meter long and of one square millimeter in section is equal to $0\cdot040 \cdot 10^{-10}$ reciprocal Siemens units at 18°. The conductivity of a liter of this water between electrodes one centimeter apart would be 10^7 greater or $0\cdot040 \cdot 10^{-3}$. If there were present in the water a gram-equivalent of H and of OH ions, the conductivity would be equal to 455 reciprocal Siemens units, for we know from our previous considerations that a gram-equivalent of hydrogen ions between two electrodes one centimeter apart would show a conductivity of 290, while for a

---

[1] *Zeitschr. physik. Chem.* xiv. 317, 1894.

gram-equivalent of OH ions it would be 165. If then 455 were found as the conductivity of the water, the solution would be normal as regards the H and OH ions. Instead of 455, $0.040 \cdot 10^{-3}$ has been found, therefore the concentration of these ions is $\dfrac{0.040 \cdot 10^{-3}}{455}$ or $0.9 \cdot 10^{-7}$ normal—that is to say, 1 gram of hydrogen and 17 of hydroxyl ions are present in about eleven million liters of water.

Supersaturated Solutions.—The idea has been prevalent for a very long time, and has not even yet disappeared, that supersaturated solutions have a characteristically different behaviour from saturated and unsaturated solutions. The conductivity measurements have proved, however, that the supersaturated solutions manifest no peculiarities not also possessed by the other solutions. If, for example, the conductivity of the solution of a salt, whose solubility increases rapidly with rise of temperature, be measured at different temperatures so chosen that at the lower ones the solution shall be supersaturated, while at the higher it is not, it will be found, by arranging the results in a co-ordinate system, that the change of the conductivity with the temperature has always been regular, and that no irregularities in the curve are present to indicate a change in the nature of the solution on passing into the supersaturated condition.

Temperature - Coefficient.—According to the experiments of Kohlrausch, the change of conductivity is a linear function of the temperature, and between wide limits may be obtained from the formula $\lambda_t = \lambda_{18}(1 + \beta(t - 18))$. In this formula λ_t and λ_{18} are the conductivities at the temperature $t°$ and $18°$ respectively, and β is the temperature-coefficient, $18°$ instead of $0°$

being taken as the starting-point. β is therefore given
by the formula

$$\beta = \frac{\lambda_t - \lambda_{18}}{\lambda_{18}(t - 18)} .$$

It has been found to be true that with good conducting
eloctrolytes the temperature-coefficient is greater for
those whose equivalent conductivities are small, and
less for those whose conductivities are great. The
actual difference in magnitude of the temperature-
coefficients for solutions of different electrolytes are not
usually very great. For most dilute, strongly dissociating salt solutions it amounts to about $0\cdot025$ at $18°$.
In other words, the conductivity in such a case would
change by about $2\cdot5$ per cent for a temperature-
difference of one degree. This shows the necessity of
making conductivity measurements at constant temperatures.

. If we imagine that the ions in their motion through
the water have to overcome a certain resistance, we
comprehend why there exists a parallelism between
the changes of the viscosity and the electrical conductivity of many solutions with changes of temperature. Strict proportionality does not, however, exist
between the two.

Finally, with regard to the temperature-coefficient
it is worthy of observation that negative values are
by no means uncommon—that is, the conductivity
sometimes decreases with rise of temperature. The
conductivity of a solution is dependent upon the
number of the ions and upon their rates of motion.
It is evident that these rates depend upon the
friction experienced by the ions in the water, and
since the internal friction of water diminishes with

rising temperature, we can assume that the friction of the ions would also decrease, especially in salt solutions, where, owing to the already highly dissociated state, a change in dissociation can only be of small value, accompanied by a corresponding increase in the conductivity. A diminution in conductivity with rise of temperature is then only conceivable through the assumption of such a simultaneous decrease in the number of ions; in other words, the dissociation decreases, so that the influence of the decreased friction is over-compensated. To many this assumption may seem unjustifiable in that it is in contradiction with conclusions drawn from the kinetic theory of gases, which suggests that with rising temperature an increase in dissociation always takes place. From the mechanical theory of heat we know that this conclusion is erroneous. From this theory we have the following general rule—*If one of the factors determining the equilibrium of a system is varied, the state of equilibrium undergoes a change in that direction which tends to counteract the original variation of the factor.*

Suppose that at a certain temperature a saturated solution of a substance is in contact with an excess of that substance. On warming the solution, that change takes place which is accompanied by a cooling: if the salt dissolves with absorption of heat, more salt will enter the solution; if with generation of heat, salt is precipitated. According to this method of reasoning, all electrolytes which tend to associate on raising the temperature, and consequently all those possessing negative temperature - coefficients of conductivity, must be characterised by negative heats of dissociation, it being understood that the heat-evolution occurring on combination of two ions to form an undissociated molecule is the heat of dissociation, and that the quantity of heat communicated to the surroundings is to be considered positive, that absorbed negative.

There is one possible method of testing the accuracy of this conclusion, and that is the determination of the heat of dissociation itself.

Heat of Dissociation.—According to the dissociation theory, the process of neutralisation of a strong base with a strong acid is nothing but the association of the H ions of the acid with the OH of the base to form undissociated water. We have already learned that the degree of dissociation of water is very slight—that is, the product of the H ions and the OH ions is extremely small. When H and OH ions come into contact in a solution, the equilibrium between their product and the quantity of undissociated water must, in accordance with the laws of mass-action, attain a certain value determined by the characteristic dissociation constant of water. We may consider the quantity of undissociated water in a solution as constant compared with the quantity of ions, since a change in consequence of the magnitude of this quantity is usually immeasurably small. Thus we are practically correct in assuming that all H and OH ions entering the water disappear, as the value of the ion product of the pure water cannot be altered. Before mixing an alkali and an acid solution we have in the one X and OH ions, and in the other H and Y ions; after mixing, there are X and Y ions, which compose the usually highly dissociated salt. The acid and the basic radical play no *rôle* in the neutralisation, therefore the heats of neutralisation of all highly dissociated acids and bases must be equal, and *their value, 13520 cal. (for 21·5°), really represents the heat of dissociation of water*—that is to say, by the union of one gram-equivalent of H with one of OH ions to form undissociated water, 13520 cal. are set free. This heat of dissociation has nothing to do with the heat evolved through the union of gaseous hydrogen and oxygen to form water.

If we neutralise a partially dissociated acid with a highly dissociated base, the heat evolved will be dependent not only on the heat of dissociation of the water, but also on the heat of dissociation of the acid. If N be this heat of neutralisation, x the degree of dissociation of the acid, and d the heat of dissociation per gram-equivalent, then $N = 13520 - (1 - x)d$ cal., and consequently $d = \dfrac{13520 - N}{1 - x}$ cal. All dissociating acids which exhibit a greater heat of neutralisation than 13520 cal. must have negative heats of dissociation. It has actually been demonstrated by the investigation of Arrhenius [1] that all acids possessing negative temperature-coefficients of conductivity have also negative heats of dissociation.

Isohydric Solutions.—If we determine the conductivities of two solutions, and afterwards mix equal volumes of them together, the conductivity of the resulting mixture will not in general, under the same circumstances, be the arithmetical mean of the two single values, unless we are dealing with completely dissociated substances. On mixing solutions of sodium chloride and potassium nitrate some undissociated potassium chloride and sodium nitrate must result, whereby the relations are complicated.

Bender called such solutions which do not influence one another's conductivity " corresponding solutions." Arrhenius, who specially studied the acids, called them

---

[1] *Zeitschr. physik. Chem.* iv. 96, 1889.

" isohydric." We shall here briefly consider the isohydrism of acid solutions, or more generally such solutions as contain an ion common to both. In this case two solutions are isohydric when the concentrations of the common ion are identical, for then no change in the degree of dissociation can take place on mixing, as may be seen from the following consideration. Suppose the one solution to be acetic acid and the other sodium acetate, then the equation $\dfrac{c^2}{c_1} = k$ holds for the acetic acid and $\dfrac{c'^2}{c'_1} = k'$ for the sodium acetate. Furthermore $c = c'$ since the concentration of the common CHCOO ions in both are the same. If, for example, one liter of the acid be mixed with four of the salt solution, the concentration of the CH_3COO ion cannot change (leaving out of account the slight change in volume of the solutions resulting from the mixing). The concentration of the H ions and undissociated acid is diminished to $\frac{1}{5}$, that of the Na ions and undissociated salt to $\frac{4}{5}$ of the former values, and by introducing these new concentration values into the equations we obtain

$$\frac{c \times \dfrac{c}{5}}{\dfrac{c_1}{5}} = \frac{c^2}{c_1} = k, \text{ and } \frac{c' \times \dfrac{4c'}{5}}{\dfrac{4c'_1}{5}} = \frac{c'^2}{c'_1} = k';$$

that is, a change in the degree of dissociation does not occur, the requirements of the conditions of equilibrium being fulfilled. It is evident that on mixing two such solutions the relative volumes have no influence on the result, and also that if two solutions are isohydric with a third, they are isohydric with each other.

Dielectric Constants and Dissociating Power of Various Solvents.—According to Nernst[1] there is a connection between the dielectric constant of a liquid and its dissociating power, both varying together in the same direction. An idea of the dielectric constants may be obtained from the following consideration. Imagine a metallic plate kept at a constant potential by connection with the pole of a constant element, the other pole of which is connected with the earth. Suppose at a definite distance from this first plate a second, which, being connected with the earth, is at the potential zero. In such a case the quantities of electricity, or the charges which the plates contain, is dependent upon the medium, *i.e.* the dielectric, which separates them. The magnitudes of the electrically opposite, equivalent charges, or, in other words, the capacities of the condenser when a specific medium is used, give the dielectric constant directly, when that of air is taken as unity. For water this dielectric constant is relatively very great (79·6 at 18°), for ethyl alcohol it is 25·8, for ethyl ether 4·25, and for carbon disulphide 2·6. The great dissociating power of water as compared with other substances agrees with these figures.

Degree of Dissociation. Reactive Power of Electrolytes. — As already shown under electrical conductivity, different substances exhibit very various degrees of dissociation in water as well as in other solvents, and the question naturally arises whether there exists any regularity in these different dissociation-values, whether, for example, the degree of dissociation is an additive property—that is to say, whether a definite atom or atom group always passes into the

[1] *Zeitschr. physik. Chem.* xi. 220, 1893.

ionic state with exactly the same tendency or force.
If this were so, the dissociation degrees of all electrolytes with different negative but the same positive
ions would form a series parallel to any other series
where a different positive ion was chosen.   In reality
this is not the case.   For example, almost all salts
containing ions of the same valency are dissociated to
about the same extent, but the corresponding acids and
bases (the compounds of the same radicals with hydrogen and hydroxyl) exhibit extreme differences in the
degree of their dissociation.   The halogen compounds
of zinc, cadmium, and mercury are, moreover, exceptions
to the general principle that all analogous salts are
about equally dissociated, while the other salts of these
metals dissociate in accordance with that principle.
No other law governing the dissociating tendency has
been established.   We know that, in general, salts are
highly dissociated in aqueous solutions, while, as just
stated, acids and bases exhibit every possible degree
of dissociation.   Other substances show practically no
conductivity when in solution.   It is remarkable that
pure substances, at ordinary temperatures, show little
or no conductivity.   For example, the pure acids, as
nitric and hydrochloric in liquid form, are almost
perfect insulators.   That the chemical activity is
closely connected with the dissociation may be concluded from the fact that such substances, in their
pure condition, possess scarcely any chemical action.
It may be stated as a general principle that chemical
processes between two substances are almost instant-
aneously completed when the compounds are at least
moderately dissociated (illustrated by most of the
reactions of analytical chemistry), while in cases where
only very few or no ions are present, reactions gener-

ally proceed very slowly, if at all, at ordinary temperatures. It is on this account that in the making of many organic compounds higher temperatures have usually to be employed in order to produce the desired reaction within a moderate period of time.

Conductivity of Fused Salts.—In the fused condition pure salts are good electrolytes. According to Poincaré, their conductivities may be measured by using metallic silver electrodes and adding to the electrolyte a very small amount of that silver salt which has the same anion as the salt to be investigated. The added amount of silver salt, when this is extremely small, does not affect the actual conductivity of the fused mass, but prevents, in a manner later to be described, the existence of .polarisation, and consequently the measurement of the conductivity of the electrolyte may be carried out as in the case of conductors of the first class.

An idea of the magnitudes of the conductivities of fused salts may be obtained from the following table, which shows the molecular conductivities (in reciprocal Siemens units) at the temperatures given :

		λ
KNO_3	350°	42·2
$NaNO_3$	350°	64·0
$AgNO_3$	350°	57·3
KCl	750°	85·2
$NaCl$	750°	128·2

As will be remembered, the equivalent conductivity of a $\frac{1}{50}$ normal KCl solution at 18° is 112·2.

The mixtures of fused salts, in contradistinction to solutions, exhibit, in the few cases yet investigated, conductivities which are approximately the sums of the conductivities of the component salts. Many salts possess not inconsiderable conductivities below their melting-points as well as above. Graetz has shown, from experiments on this point, that a sudden change in the conductivity of a salt at its melting-point does not exist. On the other hand, the temperature-coefficient of conductivity seems usually to possess a maximum value near the melting-point.

Absolute Velocity of the Ions.—The actual rates of motion of different ions in $\frac{cm.}{sec.}$, when they are under the influence of a certain difference of potential, have been calculated by E. Budde and F. Kohlrausch. For simplicity let us again imagine a gram-molecule of negative and of positive ion between two parallel platinum electrodes which are one centimeter apart. Suppose the difference of potential between the electrodes to be one volt. If exactly 48270 coulombs passed in the unit of time, and the anion and cathion had the same rate of motion, they would each move $\frac{1}{4}$ cm. in the time-unit, or their velocities would be a $\frac{1}{4}$, since the passage of 48270 coulombs through the section means that at each electrode half a gram-equivalent of ions separates. Through any cross-section of the electrolyte half a gram-equivalent of ions in all must pass, and therefore $\frac{1}{4}$ gram-equivalent of the positive and negative ions. In other words, an ion which at the beginning of the electrolysis was $\frac{1}{4}$ cm. from its proper electrode must cover exactly this distance in the given time, and this is its velocity.[1]

---

[1] It may seem impossible that while only $\frac{1}{4}$ gram-equivalent of ions has come towards the electrode through its motion, half an equivalent has separated there. We can, however, imagine that under certain conditions, even when no excess of the ions concerned is present at the electrodes, the failing positive and negative ions are supplied by the water. A further consideration of this point will be deferred.

The positive and negative ions together moved $\frac{1}{2}$ cm. in unit-time:

$$\frac{48270}{96540} = \tfrac{1}{2}.$$

The quantity of electricity which has passed in unit-time—that is, the current-strength C in ampères—when divided by 96540, under the above conditions, gives the velocity of the ions in $\frac{cm.}{sec.}$

If 96540 coulombs had passed through the solution, the ions would evidently have traversed one centimeter. The current-strength is $C = \frac{\text{Potential-fall}}{\text{Resistance}}$, $\frac{1}{\text{Resistance}} =$ Conductivity, therefore, since the fall in potential is here one volt, C is the conductivity expressed in ohms. Here the conductivity is the equivalent conductivity λ, and consequently $\frac{\lambda}{96540}$ represents the velocity; λ must here be expressed in ohms, as before stated. If the two ions have different velocities of migration they share in the total motion in proportion to their velocities.

Potassium chloride in 0·0001 normal solution exhibits at 18° the equivalent conductivity 128·9 ohms, therefore the total rate of motion of the ions is

$$\frac{128·9}{96540} = 0·001335,$$

and they share in the conductivity in the ratio 49 : 51. The potassium ion in a 0·0001 normal solution must actually move at the rate of 0·000654 cm. per second when the potential-fall is one volt. The corresponding distance for the chlorine ion is 0·000681 cm.

The following absolute velocities at 18° calculated for infinitely diluted solution are given by Kohlrausch :

K	= 0·00066 cm.	H	= 0·00320 cm.
NH$_4$	= 0·00066 „	Cl	= 0·00069 „
Na	= 0·00045 „	NO$_3$	= 0·00064 „
Li	= 0·00036 „	ClO$_3$	= 0·00057 „.
Ag	= 0·00057 „	OH	= 0·00181 „

It is of interest to note that a direct experimental proof of these calculated values is possible, and has been carefully executed by Whetham [1] according to a method given by Lodge. The values found by the latter for the velocity of the H ion agreed roughly with those of Kohlrausch, and were obtained in the following manner. An acid solution was brought into contact with a solution of potassium chloride in gelatine which had been reddened by sodium-phenol-phthalein. Under the action of the electric current the hydrogen ions gradually penetrated the gelatine solution and caused decolorisation, the rate of this change being observed. Whetham improved the method and determined the velocities of the copper, chlorine, and bichromate (Cr_2O_7) ions. The principle involved may be gathered from the following quotation :

" Let us consider the surface between two differently coloured salt solutions differing but little in density, and having a common colourless ion. If an electric current be passed through their contact-surface and we represent the two salts by AC and BC, the C ions move in one direction and the A and B ions in the other. If A and B are the cathions, the surface separating the two colours will move in the direction of the current, and the rate of this motion shows in every case the velocity of the ions causing the change of the colour." The observed values corroborated those calculated.

Determination of Solubility by Means of Conductivity. —In closing this chapter an interesting method for determining the solubility of salts difficulty soluble in water may be described. The amount of dissolved salt may be estimated, as shown by Hollemann,[1] F. Kohlrausch, and F. Rose,[2] from measurements of the conductivities of the solutions; far more accurately, in fact, than is possible by the ordinary methods.

When solutions are so dilute that a condition of complete dissociation may be assumed, then $\lambda_v = \lambda_\infty$. When the value of λ_∞ is known, the equation

$$\lambda_\infty = k\,\frac{Va}{cb},$$

the letters having the values given them on page 95, furnishes a method of determining V, the volume in which a gram-equivalent is contained in the saturated solution, and therefore the solubility.

$$V = \lambda_\infty \frac{cb}{ka};$$

k as well as $\dfrac{cb}{a}$ are easily determined experimentally; λ_∞ can usually be calculated.

---

[1] *Zeitschr. physik. Chem.* xii. 125, 1893.
[2] *Ibid.* xii. 234, 1893.

CHAPTER VI

HAVING dealt in the previous chapters especially with the one factor of electrical energy, the quantity of electricity, the other factor, the electromotive force, will now be considered.

Measurement of Electromotive Force.—As indicated in the introduction, the electromotive force of an element may be determined by means of a delicate galvanometer through an application of Ohm's law, $C = \frac{\pi}{R}$. Evidently $R = R_1 + R_2$ where R_1 represents the internal resistance of the cell and R_2 that of the rest of the circuit, and when R_2 is made so great that R_1 is inconsiderable in comparison with it, the deflections of the galvanometer needle, caused by two different elements successively introduced into the same circuit, are in the same relation as their electromotive forces. If one of the elements used be a normal element, the electromotive force of the other is thus easily obtained in volts. If the internal resistance is not negligible compared with the external, the unknown electromotive force may be found by determining the galvanometer deflections caused by the two elements when connected first in series and secondly opposing each other. If C represent the deflection of the needle,

$$\frac{C_1}{C_2} = \frac{\pi_1 + \pi_2}{\pi_1 - \pi_2},$$

and the desired electromotive force is

$$\pi_1 = \pi_2 \frac{C_1 + C_2}{C_1 - C_2}.$$

In more general use than the above method is that of Poggendorf, called also the " compensation method."

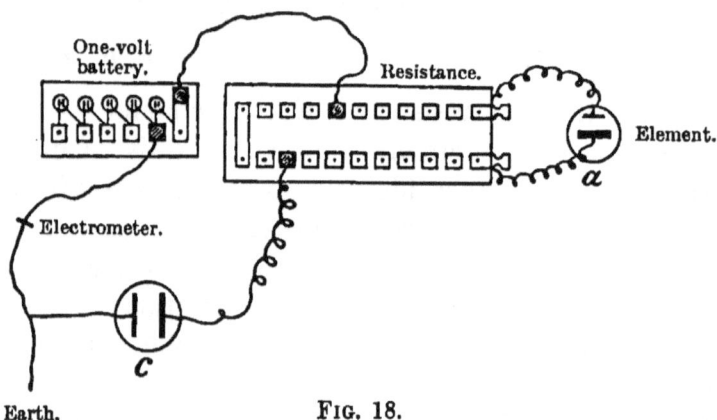

FIG. 18.

In this the unknown electromotive force is exactly compensated by a potential, the value of which is known. The following arrangement, as described by Prof. Ostwald,[1] may be advantageously employed.

The top of a wooden box (Fig. 18) is penetrated by two parallel rows of ten brass pins; of these pins the two at the extreme right are connected by good conducting plates with the binding screws, to which an element (a) is attached. The two pins at the other

[1] *Zeitschr. physik. Chem.* i. 403, 1887.

end of the box are connected together by a good conducting plate, as shown. Each pair of adjacent pins in one row is united through a resistance of 100 ohms, and in the other row through resistances of 10 ohms. These resistances consist of insulated wires soldered to the pins, and advantageously wound on glass reels placed upon the pins inside the box. There are then nine resistances of 100 and ten of 10 ohms. The two pins thus remaining unconnected are united by a thick copper wire. In the figure there is one pin too many. This could be left out, and therefore also the heavy copper wire.

When the binding screws are in connection with the poles of an element, the resistance of the circuit, exclusive of the element's internal resistance and the insignificant resistance of the connecting wire, is 1000 ohms. Throughout this resistance there is a certain regular potential-fall. Suppose this total fall to be 1 volt, then for each resistance of 100 ohms there is a potential-fall of exactly 0·1 volt, and for each 10 ohms a fall of 0·01 volt. By placing brass thimbles or caps upon the pins we may introduce between them resistances from 10 to 1000 ohms in steps of 10 ohms, and so embrace all potentials from 0·01 volt to 1 volt in steps of 0·01 volt.

The electromotive force to be measured (c) is connected with the two thimbles, which are then moved from one pin to another until compensation is reached; in other words, until the electromotive force to be measured is equal to the potential-fall between the thimbles. For the compensation of very great electromotive forces, one or more normal elements, such as the Helmholtz calomel element, having the electromotive force of one volt when prepared in a fixed manner, is used.   Any desired number of these may be connected against the electromotive force to be measured and the remainder be determined as above.

In order to determine the potential-fall distributed throughout the 1000 ohms of the resistance-box, when a certain element is in use at a, it is advantageous to proceed as follows : A normal element—for instance, a one-volt element—is inserted in the secondary circuit, which branches from the two thimbles ; the latter are then moved until the position of compensation is reached.   Suppose, in such a case, that the electromotive force of a one-volt element is exactly compensated when there are 800 ohms between the thimbles, then there is a potential-fall of one volt through 800 ohms, or 1·25 volt through 1000 ohms.   It is necessary, of course, that the normal element used should have a lower electromotive force than that at a.

A Lippmann electrometer, as arranged by Ostwald, may be used to determine when equality has been attained between the unknown electromotive force and the potential-fall by which it is compensated.   The form shown in Fig. 19 is usually sufficiently sensitive, and is described in the *Zeitschr. physik. Chem.* v. 471, 1890.

" A platinum wire, partly encased in a glass capillary, leads from an insulated binding screw and passes into the mercury at the bottom of the bulb b, which also contains a 10 per cent sulphuric acid solution.   The capillary tube c opening into b is filled in its upper part with acid ; its lower part contains mercury, as likewise the tube d, which is in connection with a second binding screw.   The position of the mercury in the capillary tube c may be regulated through altering the inclination of the capillary by means of the screw at *f.* That this apparatus may give satisfactory results, it should be short-circuited just before use, and consequently it was connected with a switch so constructed that on breaking the current the electrometer was always short-circuited and on making the current this connection within itself was destroyed. In measuring electromotive forces, so much of the resistance of the box was brought between the thimbles that the mercury in the capillary remained at rest on closing the circuit. A millimeter scale placed beneath

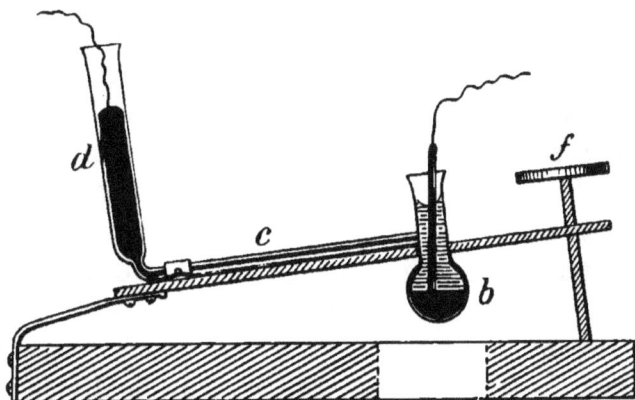

Fig. 19.

the capillary, and a lens above it, aided in the measurement. It was possible to approximately estimate a thousandth volt. One hundredth volt corresponded to $3\frac{1}{2}$ divisions of the scale."

This description suffices for the present purposes, and a study of the theory of the phenomenon will be taken up later.

The following normal elements are commonly used :—

1. The so-called Helmholtz calomel element, consisting of zinc, zinc chloride solution of 1·409 sp. gr.

at 15°, calomel, mercury. This element, when made in the prescribed manner, possesses an electromotive force of one volt[1] at about 15° C. Its change with the temperature is slight, being 0·00007 volt for 1°.

2. The Clark element, composed of zinc, a paste of zinc sulphate, a paste of mercurous sulphate, mercury, has an electromotive force of $1·434 - 0·001$ $(t - 15°)$ volt, where t is its temperature.

3. The Weston or cadmium element, composed of cadmium, a paste of cadmium sulphate, a paste of mercurous sulphate, mercury, has an electromotive force of about 1·02 volt. It is preferable to the Clark element, because its temperature-coefficient is much smaller.

Reversible and Irreversible Cells.—Any arrangement which, through chemical reaction or physical processes, such as diffusion, etc., is capable of producing electrical energy is called a galvanic cell; whether the reaction takes place between a solid and a liquid or between two liquids is of no account. Cells or elements, as they are also called, may be divided into two classes—the *reversible* and the *irreversible*. For example, the Daniell element, consisting of zinc, zinc

---

[1] Until recently the *legal* ohm (1·060 Siemens unit) was used instead of the so-called *international* ohm (1·063 Siemens unit). It is therefore necessary to distinguish between the *international* and *legal* volts in order that the relation represented by $\dfrac{\text{volt}}{\text{ohm}} = \text{ampère}$ remain intact; the latter is about 0·3 per cent less than the former. In scientific treatises both units are in use, and it is not uncommon to find that, through the use of the international unit in the calculations and of the legal volt in the measurements, mistakes are made. In the following theoretical considerations the international volt is the unit used. In those illustrations which are taken from other writers' work it is not stated which unit is employed, but the experimental errors are usually greater than the differences which are introduced by the change of the unit.

sulphate solution, copper sulphate solution, copper, is classed with the former.

Imagine the electromotive force of a Daniell element exactly compensated by a second electromotive force. Diminishing the latter a little, the Daniell element becomes active, zinc goes into solution, and copper precipitates. Increasing the opposing electromotive force so that it is greater than that of the Daniell, the copper redissolves and zinc is precipitated; thus the cell will exactly assume its previous condition. Of a reversible cell it is theoretically true that, at constant temperature, the maximum electrical energy which can be obtained through its action exactly suffices to bring the cell back to its former condition. This is at the same time the definition of a reversible cell.

An example of an irreversible cell is that first given by Volta, consisting of zinc, dilute sulphuric acid, silver. When this cell is active, zinc dissolves and hydrogen separates at the silver electrode, and is evolved. From the latter fact it is evident that the original condition cannot be reproduced by reversing the current; on the contrary, silver goes into solution, and hydrogen separates at the zinc electrode.

A characteristic of the reversible elements is that when the current strength is not too great, the electromotive force which they possess immediately after becoming active, remains nearly constant as long as material necessary to the chemical reaction is present. On the other hand, in the irreversible cells, the initial high electromotive force falls considerably, and reaches a nearly constant minimum only after some time. Hence the terms *polarisable* and *unpolarisable*. More definite information regarding this point will be given in the chapter on polarisation. It may be here stated that a metal dipping into a solution which contains a sufficient number of its own ions, is an unpolarisable electrode. In the Daniell cell both electrodes are unpolarisable, and consequently the whole cell.

Since the present condition of the science renders a clear insight into the characteristics of reversible cells essential, our attention may now advantageously be devoted to them.

Relation between Chemical and Electrical Energy II.—As already known, shortly after the discovery of galvanism, Volta advanced the hypothesis that the principal source of electromotive force was the point of contact between different metals, and he did not consider it impossible to make a cell consisting only of metals, and thereby produce perpetual motion. The law of the conservation of energy had not then been clearly defined, and the feeling of the necessity for a logical cause of a phenomenon was not always present.

The assumption of Volta was later altered, so that the electrical energy produced was considered as derived from the chemical reactions taking place at the surfaces of contact between electrode and liquid. To the points of contact between the metals, however, the production of considerable potentials was still accredited in accordance with the former assumption. It certainly seems as though even a superficial consideration would lead an unbiassed mind to find something both remarkable and improbable in the production of the electrical energy at one point, and the chief potential difference at another. In fact, there was no longer any reason for imagining the production of any considerable potential differences between the metals in the circuit. According to our present knowledge, these possible potential differences between the metals amount, at the most, to but a few hundredths of a volt. We now consider the principal potential difference to be at that point where the electrical energy is produced, and are thus able to explain satisfactorily the existing relations.

The question now arises: How may the amount of electrical energy which an element is capable of producing be calculated from the known chemical energy —or better—from the heat-effects of the reactions, since the latter constitute the measure of the chemical energy? We have seen in the introduction that the assumption originally made by Helmholtz and William Thomson, that the quantities of heat concerned changed completely into electrical energy, is untenable. It is only in certain rare cases that this simple condition exists. About twenty years ago Gibbs, Braun, and von Helmholtz succeeded in determining the existing relations by means of calculation.

The first law of energy is: *Energy cannot be created nor destroyed,* i.e. *the total amount of energy is constant.* This does not, however, preclude the possibility of the transformation of one kind of energy into another. It is the second law which deals especially with this point. This may be enunciated in many ways. It is thus expressed by Clausius: " Heat cannot pass of itself from a lower to a higher temperature." The general statement of Nernst expresses the same thing in a slightly different way which is preferable to the above: " *Every process which takes place of itself (that is, without external aid), and only such a process, is capable of doing a certain definite amount of external work.* This principle must be considered as a conclusion drawn from experience. Conversely also we may deduce the principle that an application of external work is necessary to cause a process which takes place of itself to proceed in an opposite direction." Accordingly work is necessary in order to bring heat from a lower to a higher temperature since the reverse process takes place of itself.

If we allow a body to change of itself isothermally from a condition A into another B—that is, in such a manner that the temperature remains constant—the maximum amount of external work which can be obtained is always the same, whatever may be the way in which the process is completed, whether it be osmotically, electrically, or otherwise. Knowing the maximum work obtainable in a certain way, *e.g.* osmotically, the quantity of electrical energy is also known. If the quantity of material be known, then from Faraday's law the electromotive force or intensity of this electrical energy is determined because $\pi = \dfrac{\text{Energy}}{\text{Quantity of Electricity}}$. It is evidently the maximum of available work which is of importance here. If any loss of work occurs, the amount remaining would be quite indeterminable by this principle.

It must be clear then how important it is, especially for the calculation of electromotive force, to know exactly the value of the maximum available external work which a process represents. This we may determine by allowing the body under consideration to change " reversibly " from one state into the other at a constant temperature. Let us take, for example, the case of a gas of volume v expanding isothermally from the pressure p to that of p_1. The maximum external work can be obtained when the pressure of the gas is almost completely compensated (*i.e.* to an infinitely small residuum), the process being also reversible at any time by the application of a pressure against that of the gas, exceeding the latter by an infinitely small amount.

Theoretically, in order to get the maximum work, a state of equilibrium must exist, and when this is not the case there is a certain amount of the available work appears in the form of heat and is lost.

A body having passed from a condition A into another, B, in a reversible manner, and having also been then reversibly returned to the condition A, has gone through a reversible cycle. We shall make use of such a reversible cycle in order to calculate the quantity of work (so important for electrochemistry) which may be performed when a certain quantity of heat passes from a higher to a lower temperature. For this purpose let us consider an ideal or perfect gas, since the calculation of the quantities of work is thereby much simplified. We must be able to determine the quantity of work obtainable when a gas of volume v and pressure p changes isothermally to a volume v_1 and pressure p_1. This amount of work is the same as would be produced if an "ideal" solution of volume v and osmotic pressure p changed isothermally to v_1 and p_1. As frequent use of the latter will be made, its derivation here is of twofold interest.

When a gram-molecule of a saturated vapour is in contact with its liquid, the volume and pressure of the former being v and p, the maximum work obtainable by the expansion of the vapour to v_1 under the constant pressure p is easily calculated. Imagine the increase of the volume v divided into infinitely small parts designated by dv, then the work obtainable during the expansion dv is pdv, and the total work is $p\int_{v}^{v_1} dv$, that is, p times the sum of these infinitely small amounts dv, from the value v to that of v_1, consequently $= p(v_1 - v)$. Attention is here called to the introduction, p. 5, where it is shown that the product pv, therefore $p(v_1 - v) = pv_2$, represents a quantity of work.

In the case under consideration the relations are not quite so simple, the pressure not remaining constant, but changing on the other hand with change of volume, until it reaches p_1. It is not enough then merely to add together the values dv; the sum of the endless number of infinitely small amounts of work pdv must be known, where the value of p is not a constant but a function of v, that is, in this case always possessing a value dependent upon that of the corresponding v. It may be expressed

$$A = \int_{v}^{v_1} pdv.$$

The values p and v are dependent upon each other in a definite and known manner. For molecular quantities we obtain from the gas-equation $pv = RT$, $p = \dfrac{RT}{v}$. Substituting this value of p in the above equation and placing the constants before the sign of summation (the integral sign), we get

$$A = RT \int_{v}^{v_1} \frac{dv}{v}.$$

There is here only one variable, and the integral is determinable. We know that

$$\int_v^{v_1} \frac{dv}{v} = \ln \frac{v_1}{v}, \text{ or } = \frac{1}{0\cdot4343} \log \frac{v_1}{v},$$

where ln signifies the natural and log the ordinary logarithms; consequently

$$A = RT \ln \frac{v_1}{v} = \frac{RT}{0\cdot4343} \log \frac{v_1}{v}.$$

Since $\frac{p}{p_1} = \frac{v_1}{v}$ according to the Boyle-Mariotte law, we have

$$A = RT \ln \frac{p}{p_1} = \frac{RT}{0\cdot4343} \log \frac{p}{p_1}.$$

A graphical method [1] may be advantageously employed to make the calculation simpler to those who find difficulty in understanding the above expressions of higher mathematics.

In a rectangular co-ordinate system measuring values of p on the axis of ordinates and v on the axis of abscissæ, and using the values of p and v obtained from the gas-equation $pv = RT$ as applied to a given gas, we obtain a curve as shown, which is a right-angled hyperbola, the equation of such a hyperbola being $xy = $ constant where x and y are the rectangular co-ordinates.

Suppose the values of a point a on the curve are p and v, while those for β are p_1 and v_1. Allowing the gas to change isothermally from the condition of a to that of β, the value of the work done by the gas is represented by the area a, β, γ, δ expressed in gram-centi-

---

[1] Ostwald, *Grundriss d. Allgem. Chemie*, p. 71.

meters, p being measured in grams and v in centimeters. The magnitude of the quantity represented by this area may be approximately obtained in the following elementary manner. Imagine the gas starting from the condition represented by a changed slightly so that it is in the condition a', its pressure and volume now being p' and v'. The area a, a', δ', δ represents the available work as formerly,

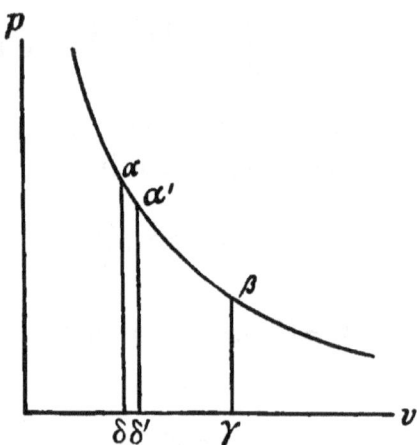

FIG. 20.

and this is nearly $\frac{p+p'}{2} (v - v')$. Proceeding in this manner we obtain the larger area a, β, γ, δ as the sum of many small areas, and the corresponding work as the sum of the many small corresponding quantities of work. The exact expression for the work cannot be obtained in an elementary manner, but, derived as above, has the value

$$\frac{RT}{0\cdot4343} \log \frac{p}{p_1}.$$

In this expression it is evident that the available work is proportional to the absolute temperature of the gas, and further, that it does not depend upon the absolute values of the pressure or volume, but upon the relation between the respective values of each. Accordingly the amount of available work is, for example, the same whether a gas passes from a pressure of ten atmospheres to one, or from one atmosphere to one-tenth. It may be recalled that if it is desired to express A in mean gram-calories, the value for R is $1\cdot96$; if expressed in gram-centimeters, $R = 84700$.

If a gas expands so that its pressure is diminished to a hundredth atmosphere, or its volume has become one hundred times greater, the maximum work obtainable in the process at $T = 290°$ ($17°$ C.) is

$$A = \frac{1\cdot96 \times 290}{0\cdot4343} \log \frac{100}{1} \text{ gm.-cal.} = 2617 \text{ gm.-cal.}$$

also

$$A = \frac{84700 \times 290}{0\cdot4343} \log \frac{100}{1} \text{ gm.-cm.} = 113120000 \text{ gm.-cm.}$$

It may be well to remark that this work which is obtained in the isothermal expansion of the gas is not taken from the internal energy of the gas itself, but the corresponding quantity of heat is extracted from the surroundings. The gas only serves as a medium for the transformation of heat into work (p. 4).

The previously described cyclical process may now be considered and the quantities of work or heat there coming into play calculated.[1]

One gram-molecule of a gas is compressed from volume v_1 to v at constant temperature. The work which is necessary to do this is

$$A = RT \ln \frac{v_1}{v}.$$

This work is converted into heat, which is absorbed by the surroundings, and the quantity of heat thus set

[1] The demonstration is here given as in Nernst's *Theoretische Chemie.*

free must be equivalent to the work done, according to the first law of energy, or

$$W = RT \ln \frac{v_1}{v}.$$

The gas is now brought into surroundings of temperature $T + dT$. The quantity of heat (m) thereby absorbed by the gas is negligibly small as compared with W (moreover, the same quantity is later given out). The volume v being kept constant during the change of temperature, no external work is done. If now the gas be allowed to expand from v to v_1, the work

$$A_1 = R(T + dT) \ln \frac{v_1}{v} = RT \ln \frac{v_1}{v} + RdT \ln \frac{v_1}{v}$$

may be obtained. The equivalent quantity of heat is taken from the surroundings:

$$W_1 = RT \ln \frac{v_1}{v} + RdT \ln \frac{v_1}{v}.$$

The gas is now brought into surroundings of temperature T. After the same negligible quantity of heat (m) as above has been given up by the gas, it is in its original condition.

On consideration of the whole result it is found that the quantity of work

$$A_1 - A = RdT \ln \frac{v_1}{v} = W \frac{dT}{T}$$

has been obtained. The equivalent amount of heat

$$W_1 - W = RdT \ln \frac{v_1}{v} = W \frac{dT}{T}$$

has been transformed into work, but at the same time the quantity of heat

$$RT \ln \frac{v_1}{v}$$

has disappeared at the temperature $T + dT$, and been recovered at the temperature T. In other words: In the fall of the quantity of heat $RT \ln \frac{v_1}{v} = W$ from $T + dT$ to T, the heat represented by

$$W \frac{dT}{T}$$

has been changed into work.

This is a general result. It is always true that when any amount of heat x is brought from a high to a lower temperature, the maximum amount which can be changed into work is represented by

$$x \frac{dT}{T},$$

where dT is the change in temperature.

To aid the comprehension the following remarks may be of use. The passage of heat from higher to lower temperature may be compared with the parallel case of the passage of electrical energy from higher to lower tension. The quantity of electrical energy $\epsilon\pi$ may be changed to $2\epsilon\frac{\pi}{2}$, that is, the total quantity of energy remains unaltered on transformation, the two factors simply changing their values in inverse proportion. The temperature T is the intensity-factor of heat-energy (Q), accordingly $Q = xT$, x being the unknown capacity-factor. Since $x = \frac{Q}{T}$, for Q the form $\frac{Q}{T} T$ is obtainable. Heat at a temperature of $100°$ may be expressed by $\frac{Q}{100°} 100°$, heat at $50°$ by $\frac{Q}{50°} 50°$. The capacity-factor is double its previous value, while the intensity-factor is one-half. $\frac{Q}{T}$ is called the *entropy*, and it is evident from the formula that its magnitude is greater the lower the temperature. Entropy tends towards a maximum.

The difference between the heat and the free transformable energy lies in the fact that the transformation in the case of the latter may theoretically take place in either direction without the use of work, while in the former a change from lower to higher temperature can only occur through consumption of work.

Let us apply these considerations to the reversible galvanic elements. If the heat evolved by the reactions taking place within such an element having no internal resistance, be entirely changed into electrical energy while the element is immersed in a calorimeter, no heating effect would be observed. The reason is that just as much energy as was produced would be consumed as electrical energy (capable of transformation into work) in the external circuit. As a matter of fact this simple relation very seldom exists, and therefore a generation of heat in the calorimeter can usually be observed.

Imagine a reversible cell of electromotive force π at the temperature T, and suppose the quantity of electricity 96540 coulombs or ϵ_0 be passed through it, then the maximum electrical energy which may be produced is $\epsilon_0\pi$. Let Q be the sum of the heats of the corresponding reactions. The action of the cell is attended by absorption of heat, the heat absorbed being $\epsilon_0\pi - Q$, according to the first law of energy. Suppose the temperature increased by dT and the amount of electricity ϵ_0 again sent through the cell, but in the opposite direction, and under the new electromotive force, $\pi + d\pi$; the amount of work thus consumed will be $\epsilon_0(\pi + d\pi)$. The corresponding sum of the heats of reaction in this reversed process has changed but little, and, neglecting this change, is Q. The heat generated in the element is in this case equal to the difference between the electrical energy used and the heat taken up in the chemical processes, and is thus equal to $\epsilon_0\pi + \epsilon_0 d\pi - Q$. If the element be brought again to the temperature T, it is once more in its original condition.

As the end-result of the process, the work $\epsilon_0 d\pi$ has been performed, and accordingly the equivalent amount of heat $\epsilon_0 d\pi$ produced. At the temperature T the heat $\epsilon_0\pi - Q$ has been lost, but at $T + dT$ the heat $\epsilon_0\pi + \epsilon_0 d\pi - Q$ has been obtained. As $\epsilon_0 d\pi$ is derived from the work done, the amount of heat $\epsilon_0\pi - Q$ has been raised from the temperature T to $T + dT$. Conversely, in order to change the quantity of heat $\epsilon_0 d\pi$ into work, the amount of heat $\epsilon_0\pi - Q$ must fall from the temperature $T + dT$ to T, consequently the following expressions are correct in accordance with page 140 :

$$\epsilon_0 d\pi = (\epsilon_0\pi - Q)\frac{dT}{T} \qquad (1).$$

$$\epsilon_0\pi - Q = \epsilon_0 T\frac{d\pi}{dT} \qquad (2).$$

$$\pi = \frac{Q}{\epsilon_0} + T\frac{d\pi}{dT} \qquad (3).$$

Since we can calculate Q from thermochemical data, or can determine it directly, we are able, with the help of the experimentally determined temperature-coefficient of the electromotive force, to calculate the maximum electrical energy obtainable, or the electromotive force of the element. In the thermochemical data the numbers always apply to a gram-equivalent or gram-molecule, the heat generated being considered positive.

If the temperature-coefficient is positive, *i.e.* if the electromotive force increases with rise of temperature, it follows from equation (2) that $\epsilon_0 \pi$ is greater than Q : the element in activity tends to become cooler, and so takes heat from the surroundings. If, on the other hand, the temperature - coefficient is negative, $\epsilon_0 \pi$ is less than Q, and the element becomes warmer. If finally the temperature-coefficient is zero, the heat of reaction is simply and completely transformed into electrical energy, and the element itself exhibits no thermal change. This latter condition is nearly realised in the Daniell cell.

It is necessary to emphasise this fact that the heat of the chemical reactions is not a strict measure of the available electrical energy of a reversible element, although experience has shown that in many cases it enables us to estimate it approximately.

The above formula of Helmholtz has been qualitatively proven by Chapski and Gockel, and quantitatively by Jahn.[1] Several apparent contradictions, as later shown by Nernst, arose from erroneously assumed values for the heat of formation of mercury compounds.

For illustration the following values found by Jahn are given. The numbers expressing calories apply to two gram-equivalents.

[1] *Wied. Ann.* xxviii. 21, 491, 1886.

[TABLE

	E.M.F. at 0° Volts.	Change in E.M.F. for $1° = \frac{d\pi}{dT}$	Elec. Energy in Calories.	Heat of Reaction in Calories.	Heat Effect in Cell.	
					Calculated.	Found.
Cu, $CuSO_4 + 100H_2O$, ZnSO$_4$+100H$_2$O, Zn.	1·0962	+0·000034	50526	50110	− 428	− 416
Ag, AgCl, ZnCl$_2$+100H$_2$O, Zn.	1·0306	− 0·000409	47506	52170	+5148	+4660
Ag, AgNO$_3$+100H$_2$O, Pb(NO$_3$)$_2$+100H$_2$O, Pb.	0·932	..	42980	50870	+7890	+7950
Ag, AgNO$_3$+100H$_2$O, Cu(NO$_3$)$_2$+100H$_2$O, Cu.	0·458	..	21120	30040	+8920	+8920

As is evident, the agreement between the heat-value of the element as observed in the calorimeter and that calculated from the difference between the electrical energy produced by the current and the corresponding heat of reaction, is satisfactory in each case.

It may be advisable to add that electrical energy may be measured by inserting the element in a circuit, the resistance of which is so great that the internal resistance of the cell is negligible in comparison. The electrical energy being allowed to change into heat, the amount of the latter generated in the unit of time is RC^2, according to Joule's law (p. 18), where R represents the resistance of the circuit, and C the current-strength. Knowing the resistance R, and having measured the current-strength, the amount of electrical energy produced in unit time may be calculated. From this the amount of energy produced when 96540 coulombs, or twice that number, pass through the circuit may be easily determined, the choice between these numbers depending upon whether one or two gram-equivalents of the substances take part in the chemical reaction. As the internal resistance of the element itself is negligible compared to the external, the Joule's heat produced within the element is insignificant, and may be left out of consideration. The heat generated in the element and measured in the calorimeter, as previously described, has evidently nothing to do with the Joule's heat, which is a measure of the electrical energy, but is the difference between the Joule's heat and the heat of the reactions taking place in the element.

The formula previously derived enables us to determine the electromotive force of a cell from a knowledge of its temperature - coefficient and of the heat of reaction. The electromotive force of reversible elements may be determined in another manner as already indicated on page 133. Before proceeding with the calculation, we must first get a clear idea of the *electrolytic solution tension* of Nernst,[1] or, as we will call it, following Ostwald, the *electrolytic solution pressure.*

Electrolytic Solution Pressure.—The expression " vapour pressure of a substance " is one commonly understood. It signifies the tendency of a substance to enter the gaseous state. If, for example, we allow water at a certain temperature to evaporate in a long cylindrical vessel in which there is a movable air-tight piston, and if a pressure is exerted upon the piston less than the vapour pressure of the water, the piston is moved upwards and more water evaporates. A condition of equilibrium is only established when a certain definite pressure is exerted upon the piston from without. The latter will then remain stationary in whatever position it be placed as soon as equilibrium between water and vapour obtains. If the pressure on the piston be slightly increased, the

---

[1] *Zeitschr. physik. Chem.* iv. 129, 1889.

vapour will be entirely condensed to water; if, on the other hand, it be slightly diminished, all the water will be changed into vapour. The weight of the piston for equilibrium represents the vapour pressure of water at the temperature of the experiment. The "solution pressure" of a substance, for example sugar, is spoken of just as the vapour pressure, and thereby is meant its tendency to pass into the dissolved state. This pressure may be measured in the same manner as the vapour pressure. The apparatus shown in Fig. 21 may be used. At the bottom of a vessel there is an excess of the solid substance A, over which is its saturated solution B, and at C pure water. s is a semi-permeable piston, that is, one which can be penetrated by the water but not by the dissolved substance. If s be weighted, the magnitude of the load determines the direction in which the piston moves. If the load be less than the pressure derived from the dissolved particles, the "osmotic pressure," s will rise and water penetrate into B, which being thereby diluted, allows more of the substance A to dissolve. If it be greater, s sinks, water passes from B into C, and the solution becoming supersaturated, some of the solid substance separates at A. Under a certain weight the condition of equilibrium must exist and the piston remain stationary at any part of the cylinder. Evidently the relations are here exactly analogous to those of the vapour pressure of water, and the magnitude of the solution pressure of the substance at a given temperature is measured by

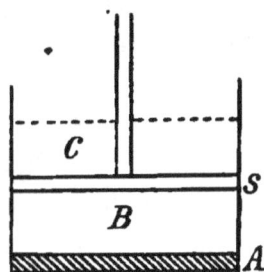

FIG. 21.

the weight of the piston when in the condition of equilibrium.

It may here be repeated that, as made evident through these considerations, the vapour pressure of water being that pressure exerted by the vapour in contact with water, that is, the "saturated" vapour, so also the "solution pressure" of a substance is the osmotic pressure of the solution which is in equilibrium with the substance, that is, the "saturated" solution.

This conception may finally be applied to the passing of substances, chiefly elements, and especially metals, into the ionic condition. Hydrogen and the metals are capable of forming only positive ions; chlorine, bromine, iodine, etc., on the contrary, form only negative ions. The magnitude of this "electrolytic solution pressure" may be conceived as determined in exactly the same manner as the ordinary solution pressure. We imagine the substance in contact with water saturated with the ions in question, under a similar piston, which separates the saturated solution from the water, and is impermeable for these ions. The equilibrium with the osmotic pressure of the ions will be brought about by a certain weight of the piston, and no ions will enter the solution from the substance nor pass out of solution. The weight of the piston in equilibrium represents the value of the electrolytic solution pressure, which is usually represented by P, and also expresses the equally great and oppositely directed osmotic pressure of the ions. This method is practically inapplicable, because in no case can appreciable amounts of positive or negative ions alone come into existence; this does not, however, affect the value of the conception.

In order to explain the production of a potential

difference through the contact of a solid substance
with a liquid, imagine a metal dipped into pure water,
and that a certain amount of metal ions is produced
owing to the electrolytic solution pressure. The metal
at the same time becomes negatively electrified, since
both kinds of electricity must be simultaneously produced whenever electrical energy comes into existence.
The solution is thus positively electrified and the metal
negatively, and there is found a so-called double layer
("Doppelschicht") of electricities of opposite signs.
The ions sent into the solution with positive charges
and the negatively charged metal attract each other;
in other words, a potential difference is produced.
The solution pressure constantly tends to send more
ions into solution, while the electrostatic attraction
of the double layer opposes this action, and evidently
equilibrium is reached when the opposing tendencies
are equal. Since the ions have very high charges of
electricity, this condition of equilibrium occurs before
weighable quantities of the ions have passed into the
water. In the case of pure water the potential
difference or strength of the double layer depends only
upon the magnitude of the solution pressure, but if
the metal be in a solution of one of its salts, another
factor is introduced, due to the metallic ions already
present. The osmotic pressure of these ions opposes
the entrance of new ions of the same kind. It may
occur that this osmotic pressure is exactly in equilibrium with the electrolytic solution pressure of the
metal, consequently the latter will yield no ions and
will not become negatively charged; in short, under
these circumstances there will be no double layer produced. The nature of the negative ions of the salt in
solution has no influence.

If the osmotic pressure of the metal ions differs from the solution pressure, two different cases may be distinguished according as the former or the latter is the greater. In the second case ions pass from the metal to the solution as in pure water, and a double layer is the result. This would evidently not be as great as in pure water, since so many ions cannot enter the solution, owing to the fact that the electrolytic solution pressure is opposed by the osmotic pressure of the ions already present. In the other case ions separate from the solution and are precipitated upon the metal communicating their positively electric charges to it. The metal thus becomes positively, the solution, which formerly contained equivalent amounts of positive and negative ions, negatively electrified, and again the electrical double layer is produced, the attraction of which opposes the previously superior osmotic pressure and adds itself to the solution pressure. This proceeds until the condition of equilibrium is reached. Here also the quantity of ions which are precipitated is unweighable. The strength of the double layer and the electrostatic attraction due to it is evidently dependent upon the osmotic pressure of the metal ions in the solution.

In all, three cases must be distinguished:

(1) When $P = p$, where P is the solution pressure and p the osmotic pressure of the metal ions considered, equilibrium exists and no potential difference or double layer is present between solution and metal.

(2) When $P > p$, the metal is negatively electrified and the solution positively. The electrostatic attraction opposes the solution pressure.

(3) Finally, when $P < p$, the metal is positively

electrified and the solution negatively. The electro-static attraction is added to the solution pressure.

On turning our attention to the actual experimental facts it is found, as will be seen later, that the alkali metals, and also zinc, cadmium, cobalt, nickel, and iron, are always negatively charged when placed in solutions of their salts; the solution pressure in these cases is so great that, owing to the limited solubility of the salts, the osmotic pressure of the metal ions can never be raised to equilibrium with the solution pressure. With the noble metals, silver, mercury, etc., the metal is usually positively electrified in solutions of its salts. The solution pressure of the metals is here slight, and it is only by employing solutions containing very few of the ions in question, i.e. such as have very low osmotic pressure due to these ions, that it is possible to have the metal negatively charged in the solution.

With such substances as produce negative ions, e.g. chlorine, there is complete analogy. If the osmotic pressure of the chlorine ions is greater than the electrolytic solution pressure, ions pass into the condition of ordinary chlorine, and the " chlorine electrode " becomes negatively charged. In the other case the electrode becomes positively charged. As a matter of fact, as far as we know, all substances which produce negative ions have high solution pressures.

So far the electrolytic solution pressure of a substance has been referred to as if it were a constant, but, just as with the vapour pressure and ordinary solution pressure, it is only constant under certain conditions, i.e. only when the temperature and the concentration of the substance in question remains unaltered.

It is well known that the vapour pressure of water changes greatly with the temperature, but that it is affected by the concentration of the water itself, and is higher the greater this concentration, may be less commonly recognised. The fact may be recalled that if two open vessels containing water at different heights be allowed to stand in a confined space, the water distils from the higher level to the lower. The water in each vessel is under the pressure of the vapour above it, and these columns of vapour differ in height by the difference between the water levels (h). Consequently the system is not in equilibrium, the tendency being for vapour to condense under the greater pressure and be generated under the lower, which process continues until the surfaces of the water in the two vessels are at the same level, or that in one of the vessels is exhausted.

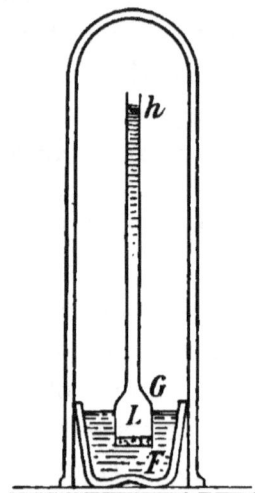

Fig. 22.

In the accompanying figure,[1] F contains pure water and L any solution, the two being separated by a membrane permeable to the water only. In the conditions represented the liquids are in osmotic equilibrium, but the vapour pressure (p_1) at the surface of the solution is less than that (p) of the water at F, and the equation $p_1 + x = p$ must represent the existing condition where x is the weight of the column of vapour, whose height is equal to the difference of level between the two liquids. If this were not true, water would distil from one surface to the other, thereby destroying

[1] *Zeitschr. physik. Chem.* iii. 115, 1889.

the existing condition of osmotic equilibrium, and would also pass through the membrane in one direction in order to reproduce the osmotic equilibrium, etc. In short, a perpetual motion would result, by which an unlimited amount of the heat of the surroundings at constant temperature could be transformed into work (through the distillation of water vapour), which is in conflict with the second law of energy.

If the upper end of the tube be closed by a membrane, allowing the passage of water vapour only, and a quantity of a gas insoluble in the liquid be placed between this membrane and the surface of the liquid, it will exert a certain pressure upon the latter, which will consequently sink to a lower level. The conditions of the equilibrium must again be that the vapour pressure (p_1') at the surface of the solution, increased by the pressure of the column of water vapour (h') between the two levels, is equal to the vapour pressure of the pure water (p), or $p_1' + x' = p$. Evidently p has remained unaltered, h' is less than h, therefore p_1' is greater than p_1, that is, at the "compressed" surface, where the water is at the greater concentration, there is a higher vapour pressure than when the water is under a lower external pressure. The increase in the vapour pressure is evidently proportional to the pressure acting on the surface.[1]

Of the ordinary solution pressure it is also known that the concentration of the substances plays an important part. This is shown by Henry's law, in accordance with which the solubility of a gas, and

[1] This conclusion was established by the work of Des Coudres and the author, which preceded the appearance of the article of Schiller on the same subject (*Wied. Ann.* liii. 396, 1894). The experiments in connection therewith were unavoidably interrupted and never concluded.

therefore its solution pressure, since the two are synonymous, is to a great extent dependent upon the pressure, in other words, upon the concentration; it is, in fact, nearly proportional to the latter.

What has been said of vapour pressure and solution pressure applies equally well to electrolytic solution pressure, and accordingly there are cells possessing certain electromotive forces dependent only upon the different concentrations of the same ion-producing substances. It is true that usually but one condition of concentration for solid substances is recognised, and consequently a single definite electrolytic solution pressure. But even here the concentration may be varied, as will be later described. The electrolytic solution pressure also varies with the temperature.

Calculation of Potential Differences by means of the Electrolytic Solution Pressure.—It is an easy matter to calculate the potential differences between an electrode and the solution with which it is in contact when the electrolytic solution pressure P of the electrode and the osmotic pressure p of the corresponding ions in the solution are known. It is evidently only the pressure of the corresponding ions which here comes into consideration; with a zinc electrode it is only necessary to know the concentration of the zinc ions in the solution. The maximum amount of work which might be obtained osmotically is determined, and considered equal to that obtainable electrically.

If a univalent element with solution pressure P is to be changed into ions of the osmotic pressure p, then the maximum work which may be obtained is equal to that obtainable by the passage of the ions from the osmotic pressure P to that of p, no work being performed by the simple change of a substance of solution pressure P into ions of equivalent osmotic pressure.   As the laws applicable to gases also hold for (dilute) solutions, the amount of work may be calculated in the same manner through replacing gas pressures by osmotic pressures.   The osmotic work of a gram-molecule is then represented by

$$RT \ln \frac{P}{p}.$$

The electrical work is $\epsilon_0 \pi$, when π represents the potential difference between electrode and electrolyte, consequently

$$\epsilon_0 \pi = RT \ln \frac{P}{p}$$

$$\pi = \frac{RT}{\epsilon_0} \ln \frac{P}{p}.$$

Obviously π is zero when $P = p$.   This agrees with the previous conclusion that in this case there is no potential difference between electrode and electrolyte.

Since the passage of one gram-ion is being considered, ϵ_0 is 96540 coulombs when the ion is univalent.   Both kinds of energy in the above equation must be expressed in the same units.   According to p. 17, $4 \cdot 24$ is the electrical equivalent of heat.   R is $1 \cdot 96$ cal.   For this reason the right side of the equation, which gives calories only, must be multiplied by $4 \cdot 24$ in order to change it into electrical units or

$$96540 \times \pi \, (\text{volts}) = \frac{4 \cdot 24 \times 1 \cdot 96}{0 \cdot 4343} \, T \log \frac{P}{p}.$$

At the temperature $17°$, T is $290°$, and

$$\pi = \frac{1{\cdot}96 \times 290 \times 4{\cdot}24}{0{\cdot}4343 \times 96540} \cdot \log \frac{P}{p} \text{ volts} = 0{\cdot}0575 \log \frac{P}{p} \text{ volts.}$$

If the ion is not univalent, then $n_e \times 96540$ coulombs would be transported with one gram-ion, where n_e is the valency. The formula thus becomes

$$\pi = \frac{0{\cdot}0575}{n_e} \log \frac{P}{p} \text{ volts.}$$

This is a fundamental equation in the theory of reversible cells.

In considering a cell composed of two metals and two solutions, as, for instance, the Daniell—zinc, zinc sulphate, copper sulphate, copper—there are four places where potential differences are produced :

1. At the point of contact between the two metals.

2. At the point of contact between the two liquids.

3 and 4. At the points of contact of both electrodes with the liquids.

The potential difference at the points of contact between the two metals is so small that it may usually be left out of account. This is also often true of that existing between the two solutions. These magnitudes will shortly be calculated. Considering only the potential differences at the points of contact of the electrodes with the liquids, the electromotive force of the cell at $17°$ is expressed by the following equation :

$$\pi = \frac{0{\cdot}0575}{n_e} \log \frac{P}{p} - \frac{0{\cdot}0575}{n_e{}'} \log \frac{P'}{p'} \cdot$$

P represents the electrolytic solution pressure of the one substance, the valency and osmotic pressure of whose ions are n_e and p. P', $n_e{}'$, and p' are the corresponding values for the other substance. The minus sign is used because at one electrode ions enter the solution, while at the other they pass from the solution; for example, in Daniell's cell zinc ions are produced, and simultaneously an equal number of copper ions separate at the other electrode, for the same number of positive and negative ions must always be present in the solution. The investigation of special cases will now be taken up.

CONCENTRATION CELLS

A. Different Concentrations of the Substances forming the Ions

1.• A cell formed of two differently concentrated amalgams of the same metal, for example, zinc, in a solution of one of its salts, as zinc sulphate, possesses, according to the previous considerations, an electromotive force at T° expressed by the formula

$$\pi = \frac{0\cdot000198}{2} T \log \frac{P}{p} - \frac{0\cdot000198}{2} T \log \frac{P'}{p} .$$

Since the concentration (p) of the zinc ions is the same throughout the solution, the formula may be simplified to

$$\pi = \frac{0\cdot000198}{2} T \log \frac{P}{P'} .$$

P and P′ are respectively the electrolytic solution pressure of the zinc in the more concentrated and more dilute amalgam. Weak amalgams may be considered as solutions in which the mercury is the solvent and, in the above case, zinc the dissolved substance. The zinc, like all dissolved substances, exerts a certain osmotic pressure which, since the amalgams are not of the same concentration, is different at the two electrodes. Since these are proportional to the concentrations, the electrolytic solution pressures of the amalgams may be assumed proportional to the osmotic pressures of the dissolved zinc.[1] From this

$$\pi = \frac{0 \cdot 000198}{2} \, T \log \frac{c}{c_1} \text{ volts,}$$

where c and c_1 are the concentrations of the zinc in the amalgams. That values of π calculated in this manner agree with those experimentally determined may be seen from the following results obtained by G. Meyer:[2]

Zinc Amalgam and Zinc Sulphate Solution

T	c	c_1	π found.	π calculated.
11·6°	0·003366	0·00011305	0·0419 volt	0·0416 volt
18·0°	,,	,,	0·0433 ,,	0·0425 ,,
12·4°	0·002280	0·0000608	0·0474 ,,	0·0445 ,, ·
60·0°	,,	,,	0·0520 ,,	0·0519 ,,

Cadmium Amalgam and Cadmium Iodide Solution

T	c	c_1	π found.	π calculated.
16·3°	0·0017705	0·00005304	0·0433 volt	0·0440 volt
60·1°	0·0017705	0·00005304	0·0562 ,,	0·0507 ,,
13·0°	0·0005937	0·00007035	0·0260 ,,	0·0262 ,,

Copper Amalgam and Copper Sulphate Solution

T	c	c_1	π found.	π calculated.
17·3°	0·0003874	0·00009587	0·01815 volt	0·0176 volt
20·8°	0·0004472	0·00016645	0·0124 ,,	0·0125 ,,

The electromotive force π of such cells can be calculated in a second way, independent of the idea of electrolytic solution pressure. The action of the

---

[1] This is equivalent to assuming that the dissolved substance is present in the mercury as atoms, which will be demonstrated from considerations of concentration cells formed from gases (p. 163).

[2] *Zeitschr. physik. Chem.* vii. 447, 1891 ; and Ostwald, *Allgem. Chem.* ii. 861.

cell consists in zinc passing from the more concentrated amalgam into the solution, and at the same time from the solution into the weaker amalgam. That is to say, zinc at an osmotic pressure p, or the proportional concentration c, changes to the osmotic pressure p_1 or the concentration c_1. The maximum amount of work thereby obtainable osmotically is

$$\frac{RT}{0\cdot4343} \log \frac{c}{c_1}$$

for a gram-atom, when the metal is assumed to be present in the mercury in the form of atoms.

The electrical value of the same process is $2 \times 96540 \times \pi$, and since the maximum amounts of work must be equal,

$$2 \times 96540 \times \pi = \frac{RT}{0\cdot4343} \log \frac{c}{c_1},$$

or

$$\pi = \frac{0\cdot000198}{2} T \log \frac{c}{c_1} \text{ volts.}$$

This is the same formula obtained by the previous method, and will also be later used in the calculation of π.

It was assumed that the metal is present in the mercury in the atomic state, and since the experimentally determined values of π agree with those calculated, this assumption may be considered justified.

If the metals had dissolved in the mercury in complexes of two atoms each, the work obtainable osmotically, through the transportation of the same amount of metal as before, would have been

$$\frac{1}{2} \frac{RT}{0\cdot4343} \log \frac{c}{c_1},$$

because the number of separate particles to be transported is only half as great. The work obtainable depends upon their number, but not upon their weight. The corresponding electrical energy would be

$$2 \times 96540 \times \pi',$$

therefore

$$2 \times 96540 \times \pi' = \frac{1}{2}\frac{RT}{0\cdot4343}\log\frac{c}{c_1}$$

and

$$\pi' = \frac{1}{2}\frac{0\cdot000198}{2}\log\frac{c}{c_1} = \frac{1}{2}\pi,$$

or in such a case the electromotive force of the cell would be only half as great as is actually found. The monatomic character of the metal molecules in mercury solutions has also been proved from measurements of the vapour-pressure diminutions.

As shown by the formula, π depends only upon the relation between the concentrations and upon the valency of the metal, and is in other respects independent of the nature of the metal.

The amalgams have been considered simply as differently concentrated zinc electrodes; it might be asked if the mercury in them does not also play the part of an electrode, and its electrolytic solution pressure come into consideration. This is not the case. If two different metals, as in a solid alloy, are in contact with the liquid, only that one is active which produces the greater electromotive force, if the amount present is not too small.

If an alloy of zinc and cadmium be placed in an acid solution, the zinc in contact with the acid dissolves first, and the solution of the cadmium only begins later. In employing such an alloy as electrode,

the greater electromotive force of the zinc is very
nearly obtained at first, and later the smaller one,
that of the cadmium. If zinc ions are present this
metal has no effect when the osmotic pressure of
these ions is so great that the cadmium dissolves
more easily. If the solution originally contains cadmium ions, a secondary reaction is introduced, which
proceeds until as many cadmium ions have been
precipitated on the electrode and been replaced by
zinc ions as is possible at the existing electromotive
force.

2. The combination mercury, a solution of mercurous salt, amalgam of a noble metal, can also be
classed as a concentration cell. From what was said
of the osmotic pressure (p. 54), it is evident that
(leaving out of account electrostriction and chemical
reactions) the volume of a liquid should be slightly
increased by the solution of a substance in it, since
the particles of dissolved substance exert an outward
pressure upon the surface of the solution. The dissolving of the substance has, in this respect, the same
effect as a reduction of the external pressure which
acts upon the liquid; the latter expands, and thereby
the concentration (*i.e.* the mass in unit volume) is
reduced. In the above-mentioned cell there are
thus two differently concentrated mercury electrodes.
Evidently only those metals may be used to dilute
the mercury whose solution pressure is weaker, since
that of the mercury only ought to come into consideration. Gold and platinum, the so-called noble
metals, adapt themselves to this end. A mercurous
salt must be used as the electrolyte. Mercuric salts
are immediately reduced in contact with metallic
mercury.

The electromotive force of this mercury concentration cell may be easily calculated, as was that of the previously described cell, either with or without the use of the idea of electrolytic solution pressure. It will be sufficient to apply the shorter method, since the electromotive force of such a cell has not yet been experimentally determined.

In the action of the cell mercury dissolves from the pure mercury electrode, where the solution pressure is greater, and is precipitated upon the amalgam electrode. The maximum work available osmotically will now be calculated and considered equivalent to the maximum available electrical work.

Suppose the pure solvent (mercury) separated from the solution (the amalgam) by a movable semi-permeable diaphragm. Let p represent the osmotic pressure of the solution, and v be the volume in which one gram-molecule of dissolved substance is contained. Let the semi-permeable diaphragm be moved under the constant pressure p in the direction of the pure solvent, until an amount of the latter equal to v enters the solution. If v be one cubic meter, for example, one cubic meter of the solvent passes through the diaphragm into the solution, and the former is moved through the volume v of one cubic meter at the constant pressure p. Let the amount of the solution be so great that the introduction of this volume v causes no appreciable change in the concentration. The maximum work which can thus be obtained is represented by the product pv, since v is the volume containing a gram-molecule of dissolved substance. But $pv = RT$ and consequently RT is the osmotic work. To obtain the equivalent electrical energy, that amount (m) of mercury gram-molecules which is contained in the volume v must pass from the one electrode to the other ; therefore

$$m \epsilon_0 \pi = \mathrm{RT},$$

and

$$\pi = \frac{\mathrm{RT}}{m \epsilon_0} .$$

The values of R, T, and ϵ_0 are known ; m is the number of gram-molecules of mercury containing one gram-molecule of the dissolved metal in the amalgam. Hence the value of π is easily reckoned.

This method serves also for determining the molecular weight of the metals dissolved in the mercury ; m is the number of gram-molecules of mercury containing one gram-molecule of the dissolved metal. By measuring π, m is obtained, and from the known concentration of the amalgam, the weight of the dissolved substance in m, which represents the molecular weight, is calculable.

3. A second mercury concentration cell is the following: Mercury under greater than atmospheric pressure, mercurous salt, mercury under atmospheric pressure. In such a cell mercury passes from the former electrode through the electrolyte to the latter. Des Coudres [1] arranged this cell as follows : A column of mercury of height h formed one electrode ; the lower end of the tube containing it, closed by means of parchment paper, was placed in a salt solution. The paper was impervious to the mercury as such, but allowed the passage of the ions. The surface of the second mercury electrode was at the same height on the parchment membrane. The height of the mercury column decreases by a definite amount when a gram-molecule of mercury passes from one electrode to the

---

[1] *Wied. Ann.* xlvi. 292, 1892.

other. The maximum work thus obtainable may be calculated, and is equivalent to the electrical energy. The work necessary for the transport of the ions through the solution may be left out of account. If 200 gms. thus leave the column of mercury, which is of great height h, the effect is the same as though 200 gms. of mercury had fallen the distance h, and the maximum available mechanical energy is $200\,h$ gm. cm. where h is expressed in centimeters. Therefore, since according to p. 17 the gm. cm. units must be divided by 10210 in order to obtain electrical units,·

$$\epsilon_0 \pi = \frac{200h}{10210},$$

and the electromotive force has the value

$$\pi = \frac{200h}{96540 \times 10210} \text{ volts.}$$

The following shows experimentally determined values compared with those calculated, and considering the difficulty of accurately measuring these small values, the agreement must be considered satisfactory.

Pressure in cm.	π calculated.	π found.
36	$7 \cdot 2 \times 10^{-6}$ volts.	$7 \cdot 4 \times 10^{-6}$ volts.
46	9·3　　,,　　,,	10·5　　,,　　,,
113	23　　,,　　,,	21　　,,　　,,

4. Finally, concentration cells may be produced from gases or aqueous solutions of different concentrations as ion-producing substance. At the first glance it may seem improbable that gases or liquids, which possess no metallic conductivity, can serve as electrodes, but through the use of a special arrangement this end

is easily reached. A platinised platinum electrode is passed from beneath into a tube closed above, the lower end of which stands in a liquid. The tube is so filled with the gas under consideration that the platinum plate is for the greater part in the gas, the remaining portion being in the liquid. The platinised platinum absorbs a certain quantity of the gas, and may be considered as a gas electrode. The only other part the platinum plays in these cells is that of conductor of the electricity. Because of its power of dissolving the gases the platinum allows of the change from the gaseous to the ionic state, and the reverse, without resistance. Such an electrode, *e.g.* one of hydrogen, belongs to the reversible class as experimentally shown by Le Blanc.[1] The quantity of work developed by the passage of a certain amount of gas into the ionic condition is exactly the quantity necessary and sufficient to produce the reverse action. Since this is true, the material of the metallic electrode can have no effect upon the electromotive force, and, in fact, equal values have been obtained with platinum and palladium electrodes.

By means of such platinised platinum electrodes reversible chlorine, bromine, and iodine electrodes may be prepared. By arranging a reversible cell of two such electrodes, using as ion-producing material the same substance for each, but in different concentrations, a concentration cell entirely analogous to that of the amalgams is the result. The electrolyte to be used is evidently one containing the same ions as the gas produces. If hydrogen be the gas, an acid is used; if oxygen (the corresponding ions of which are OH), a solution of a base must form the electrolyte. This

[1] *Zeitschr. physik. Chem.* xii. 333, 1893.

kind of a cell is independent of the nature of the electrolyte, except for the above consideration defining one of the ions.

In the calculation of the electromotive force of a gas cell, for example one consisting of two hydrogen electrodes under the pressures p and p_1, the process is the same as with the amalgam cell, except that it must be borne in mind that the hydrogen molecule contains two atoms. In the reversible change of one gram-molecule of hydrogen from the pressure p to p_1, the maximum work is represented by

$$\mathrm{RT} \ln \frac{p}{p_1}.$$

The corresponding energy, when the process is considered as an electrical one, is $2\epsilon_0 \pi$, because the molecule produces two univalent ions; therefore

$$\pi = \frac{\mathrm{RT}}{2\epsilon_0} \ln \frac{p}{p_1}.$$

The factor 2 occurs here in the denominator, even though the equation applies in this case to univalent ions.

If the calculation be made in accordance with the osmotic process, using solution pressures as on p. 156, the formula is

$$\pi = \frac{\mathrm{RT}}{\epsilon_0} \ln \frac{\mathrm{P}}{\mathrm{P}_1},$$

P and P_1 being the solution pressures corresponding to the gas pressures p and p_1 respectively. Evidently the two must be equal, or

$$\frac{\mathrm{RT}}{2\epsilon_0} \ln \frac{p}{p_1} = \frac{\mathrm{RT}}{\epsilon_0} \ln \frac{\mathrm{P}}{\mathrm{P}_1},$$

and

$$\frac{1}{2}\ln\frac{p}{p_1} = \ln\frac{P}{P_1} \ ;$$

therefore

$$\frac{p}{p_1} = \frac{P^2}{P_1^{\ 2}} .$$

That is, the squares of the solution pressures are in the same ratio as the corresponding gas pressures. This result is not difficult to understand. It may be recalled that P and P_1 represent osmotic pressures (p. 147). If the osmotic pressure P exists in a solution at the one gas electrode whose gas pressure is p, while at the other the osmotic pressure is P_1 and the gas pressure p_1, there is no potential difference at the electrodes. There is a condition of equilibrium between the gas molecules and the corresponding ions. When such a condition exists that undissociated and dissociated portions are in equilibrium, the concentration of the undissociated portion, divided by the product of the concentrations of the dissociated portions, is a constant. Moreover, the gas and osmotic pressures are proportional to the concentration, hence

$$\frac{p}{P^2} = k,$$

and also

$$\frac{p_1}{P_1^{\ 2}} = k,$$

therefore

$$\frac{p}{p_1} = \frac{P^2}{P_1^{\ 2}} .$$

Experimentally determined values of the electromotive forces of such cells have not as yet been published, so that a comparison with the calculated values is at present impossible.

The consideration of a second kind of concentration cells will now be taken up.

B. Different Concentrations of the Ions

1. The combination: silver, silver nitrate solution (concentrated), silver nitrate solution (dilute), silver, may be considered as a type of these cells. In such a cell, where the electrode furnishes positive ions, the current always flows through the cell from the dilute solution to the concentrated. Silver is dissolved in the dilute solution and precipitated from the other, this process continuing until the two solutions are of the same concentration. That the silver ions must precipitate from the more concentrated solution is evident when it is remembered that the osmotic pressure here directed against the solution pressure is greater than in the dilute solution.

Leaving out of account for the present the potential difference which exists at the point of contact between the two solutions, the electromotive force of such a cell is

$$\pi = \frac{RT}{\epsilon_0} \ln \frac{P}{p_1} - \frac{RT}{\epsilon_0} \ln \frac{P}{p},$$

where p and p_1 are the osmotic pressures of the silver ions in the concentrated and dilute solution respectively. Since the solution pressures are the same, the formula may be simplified to

$$\pi = \frac{RT}{\epsilon_0} \ln \frac{p}{p_1}.$$

This expresses the fact that the electromotive force

of such a cell is dependent only upon the relation between the osmotic pressures of the metal ions, and is independent both of the nature of the metal and of the negative ions of the electrolyte.

The electromotive force may also be ascertained by the second method, through calculating the maximum of energy represented by the osmotic change when one gram-ion of silver migrates from one electrode to the other. For this purpose the conditions of the cell before and after the electrolysis are compared.

If one gram-ion dissolves in the dilute solution, the silver concentration is increased by one gram-ion, but at the same time some silver also passes from the dilute to the concentrated solution. If n be the share of transport of the silver, n gram-ions leave the dilute solution, and the actual increase in the concentration of the latter when one gram-ion dissolves is $(1 - n)$ gram-ions. The stronger solution must evidently have its concentration reduced by this amount. A migration of NO_3 ions also takes place. If n' represent the share of transport for this ion, then $n'\, NO_3$ gram-ions pass from the concentrated to the dilute solution, since the motion is in the direction opposite to that of the silver ions. But n' is equal to $1 - n$, consequently $1 - n$ gram-ions of silver and the same number of NO_3 gram-ions move from the concentrated solution to the dilute during the passage of 96540 coulombs, i.e. from osmotic pressure p to p_1. The relation of the osmotic pressures of the p_1 anions as well as of the cathions is $\dfrac{p}{p_1}$. The work is expressed by

$$2(1 - n)\mathrm{RT} \ln \frac{p}{p_1},$$

and

$$\pi = \frac{2(1-n)\mathrm{RT}}{\epsilon_0} \ln \frac{p}{p_1}.$$

On comparing this formula for univalent metals with that obtained above, it is seen that when $n = \frac{1}{2}$, i.e. when the two ions have equal rates of migration, the formulæ become the same. When this is not the case, a potential difference exists at the point of contact between the solutions, and this requires the application of a correction to the previous formula; consequently the formula just derived is more general in its application. It will be assumed for the present that $n = \frac{1}{2}$.

The following formula is the most general one:

$$n_e \pi \epsilon_0 = n_i (1-n)\mathrm{RT} \ln \frac{p}{p_1},$$

or

$$\pi = \frac{n_i(1-n)}{n_e \epsilon_0} \mathrm{RT} \ln \frac{p}{p_1}.$$

Here n_e is the number of ϵ_0 which must be transported to cause $(1-n)$ gram-molecules of the electrolyte to pass from the concentrated to the dilute solution. The highest valency represented by the ions in a given case gives the value of n_e directly. If zinc chloride be the electrolyte, $n_e = 2$. In the concentration cell: thallium, concentrated thallium sulphate solution, dilute thallium sulphate solution, thallium, n_e is also 2. If the electrolyte be thallium nitrate, $n_e = 1$, and so on. The number of ions formed from a molecule of the electrolyte is n_i.

For dilute solutions the relation between the concentrations may be used, instead of that between the osmotic pressures. For example, for the cell: silver,

silver nitrate solution (0·01 normal), silver nitrate solution (0·001 normal), silver, 10 is substituted for $\frac{p}{p_1}$ of the formula, and the electromotive force calculated should agree closely with that measured.

Nernst [1] measured the electromotive force of the cell : silver, silver nitrate solution (0·1 normal), silver nitrate solution (0·01 normal), silver, and found $\pi = 0·055$ at 18°. From conductivity determinations it was calculated that the relation between the concentrations of the silver ions, instead of being 1 : 10, was 1 : 8·71 ; consequently

$$\pi = 0·000198 \times 291 \log 8·71 = 0·054.$$

The agreement of the experimentally found value with that calculated is evidently very satisfactory.

The following statements will serve to give a general idea of the magnitude of the numerical values. Since at 17°

$$\pi = \frac{0·0575}{n_e} \log \frac{p}{p_1} \text{ volts,}$$

it follows, where the concentrations of the ions to be considered are in the ratio 1 : 10 and the metal univalent, that

$$\pi = 0·0575 \text{ volt.}$$

If the relation of the concentrations is increased to 1 : 100 or 1 : 1000, the values of π become twice or three times as great, since π increases in logarithmic ratio.

If the ion be other than univalent, the corresponding values must be divided by the valency. Thus the cell consisting of copper and copper sulphate

---

[1] *Zeitschr. physik. Chem.* iv. 129, 1889.

solutions, in which the concentrations of the copper ions are $1:10$, would give an electromotive force of about one half that of the corresponding silver concentration cell. Measurements by Moser corroborate this statement.

The accompanying diagram graphically represents the electromotive force of concentration cells of uni-

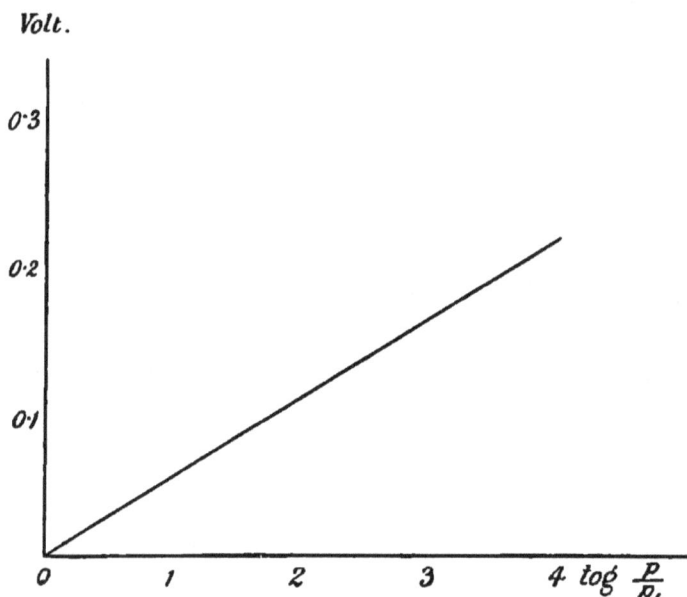

FIG. 23.

valent ions, the concentration of one solution being unity, that of the other varying from this value to $0 \cdot 0001$.

2. Another kind of concentration cell is represented by the combination : silver, silver nitrate solution, potassium chloride solution, silver coated with silver chloride. In spite of the apparent difference between this and the cell last described, the two are entirely analogous. In calculation of the electromotive force

only the osmotic pressures of the silver ions in the nitrate solution and in the solution of the silver chloride require to be taken into account. The potassium chloride is used to increase the conductivity of the silver chloride solution. In practice a solution of potassium nitrate is inserted between the potassium chloride and silver nitrate solutions, in order to present the formation of a precipitate. The formula

$$\pi = \frac{0.000198}{n_e} \, T \log \frac{p}{p_1}$$

holds good.

In calculation of π the ratio $\frac{p}{p_1}$ alone need be known. In the nitrate solution the concentration of the silver ions may be known, if a solution of a certain strength be prepared, for if not very dilute, so that complete dissociation may be assumed, the degree of dissociation may be determined. In the case of the solution of silver chloride the concentration of silver ions is not so easily ascertained. On account of the slight solubility of the chloride it is certainly very low. By means of the electrical conductivity (p. 123) the solubility in pure water may be determined, and it has thus been found that the saturated silver chloride solution at 18° is 0.0000117 normal. In such a dilute solution the salt is doubtless practically all dissociated into the ions Ag and Cl; moreover, as they are present in equivalent amounts, the solution is 0.0000117 normal for silver or for chlorine ions, and the product of these concentrations is

$$Ag \times Cl = (0.0000117)^2 = s^2.$$

Instead of a pure aqueous solution of silver

chloride, that of the cell also contains potassium chloride. From p. 87 the product of the concentrations of the ions, divided by the concentration of the undissociated molecules, is a constant independent of the dilution; and since in a saturated solution the undissociated portion must be considered constant, the same is true also of the product of the concentrations of the ions. When a relatively large amount of potassium chloride is added to a saturated aqueous silver chloride solution, the number of chlorine ions is greatly increased, and, in consequence, a certain amount of undissociated silver chloride must form and be precipitated, since the solution is already saturated with it. If c is the concentration of the silver ions after the addition, and also that of the chlorine ions derived from the silver chloride, while c_1 is the concentration of the added chlorine ions, then

$$c(c+c_1)=s^2,$$

and since c_1 is very great compared with c, the equation may be written

$$c=\frac{s^2}{c_1}.$$

To obtain the concentration of the ion corresponding to the material of the electrode, the square of the solubility (s) of the salt used is divided by the concentration of the other ion, of which an excess is added. Supposing a 0·1 normal potassium chloride solution to be used, c_1 for complete dissociation would be 0·1, but since in this concentration it is only about 85 per cent dissociated, $c_1 = 0·085$; and therefore

$$c=\frac{(0·0000117)^2}{0·085}.$$

Since the osmotic pressures are proportional to the concentrations, and the silver nitrate is 82 per cent dissociated, when the silver nitrate solution is 0·1 normal, the following holds for 18° :

$$\pi = 0 \cdot 000198 \times 291 \times \log \frac{0 \cdot 082 \times 0 \cdot 085}{(0 \cdot 0000117)^2} = 0 \cdot 44 \text{ volt.}$$

The corresponding experimentally determined value is 0·46 volt, a satisfactory agreement.

Another example of such cells is one consisting of silver, KNO_3 solution saturated with $AgBrO_3$, $KBrO_3$ solution saturated with $AgBrO_3$, silver.[1] The concentration of the silver ions in the nitrate solution is nearly the same as in pure water, since the nitrate yields neither Ag nor BrO_3 ions, and consequently has no influence on the state of dissociation of the $AgBrO_3$. The concentration of the silver ions in the potassium bromate solution may be calculated as before, from the solubility of the silver bromate in water and the concentration of the BrO_3 ions added. When the values so obtained are substituted in the formula, $\pi = 0 \cdot 0612$ volt for 0·1 normal, and $\pi = 0 \cdot 0454$ for 0·05 normal potassium bromate solution. The experimentally determined magnitudes are 0·0620 and 0·0471. The current, as before, passes in the cell from the weaker to the more concentrated solution of silver ions, or from the bromate to the nitrate solution.

Electrodes in which the metal is in contact with one of its difficultly soluble salts, and also in the presence of a solution of a soluble salt with the same negative ion, were called by Nernst electrodes of the second order, or as regards the negative ions, reversible electrodes. Ostwald showed that these are not to

[1] *Zeitschr. physik. Chem.* xiii. 577, 1894.

be distinguished from ordinary metal electrodes in a solution of one of their salts.

3. A third kind of concentration cell consists of those in which one of the electrolytes is a complex salt. As a type of this class may be mentioned the combination : silver, silver nitrate solution, potassium cyanide solution containing a little silver cyanide, silver. In the latter solution the complex salt $KAg(CN)_2$ is formed, the ions being K and $Ag(CN)_2$. This negative ion is in turn dissociated to an extremely slight extent into 2(CN) and Ag, and it is the concentration of this latter silver ion which in this solution is to be taken into account in considering the electromotive force of the cell. It is evidently somewhat dependent upon the quantity of silver cyanide. Since it is at present impossible to measure the concentration of this small quantity of ions in the solution of the complex salt, it is also impossible to calculate the electromotive force of such cells. On the other hand, the measurement of the electromotive force gives a means of calculating the concentration.

A determination of a parallel case is here given in which the electromotive force of the cell : Mercury, mercurous nitrate (0·1 normal), mercurous sulphide in sodium sulphide, mercury, was measured.[1] The value of π at 17° was found to be 1·252 volt, or

$$1 \cdot 252 = 0 \cdot 000198 \times 290 \log \frac{p}{p_1},$$

where p and p_1 represent the osmotic pressures or concentrations of the mercury ions in the nitrate and in the sodium sulphide solutions. Further,

$$\log \frac{p}{p_1} = 21\cdot 8,$$

and

$$\frac{p}{p_1} = 10^{21\cdot 8}.$$

Assuming complete dissociation, there are 20 grams of mercury ions in a liter, or 1 mg. ion in $0\cdot 00005$ liter, of the $0\cdot 1$ normal mercurous nitrate solution. This latter number, multiplied by $10^{21\cdot 8}$, gives the number of liters of the sodium sulphide solution containing one milligram of mercury ions.

A means of determining the solubility of the difficultly soluble salts, and thereby the ion concentration, has already been observed in the electrical conductivity. These considerations furnish, however, a second method far surpassing the first in delicacy. In fact, it is exactly at those extremely low concentrations, where all other methods are without avail, that the advantages of this one are most prominent, since the electromotive force becomes higher the greater the difference in the concentrations. In those cases where determinations by both methods have been possible, satisfactorily agreeing results were obtained.

Attention may be called to the following important fact. In the three cells: (1) Silver, $0\cdot 1$ normal silver nitrate, $0\cdot 1$ normal potassium chloride with silver chloride, silver; (2) Silver, $0\cdot 1$ normal silver nitrate, $0\cdot 1$ normal potassium bromide with silver bromide, silver; (3) Silver, $0\cdot 1$ normal silver nitrate, $0\cdot 1$ normal potassium iodide with silver iodide, silver, the electromotive force increases from the first to the third. This is a consequence of the fact that the silver chloride is more soluble than the bromide, and this in turn more soluble than the iodide. In such cells as these the electromotive force is greater the less soluble the salt. With the complex instead of the insoluble salts, as is illustrated by the 0·1 normal potassium cyanide solution, to which some silver cyanide was added, the electromotive force is greater the fewer the metal ions furnished by the salt (in this case silver). If a series of such cells be arranged in the order of their electromotive forces, beginning with the lowest, the order is also that of the solubility, or, in the last-mentioned case, of the decomposition.

Each salt in the series will dissolve in, or else react with, any of the saturated solutions of the cells following in the series. For example, silver chloride added to the potassium bromide solution forms silver bromide; silver bromide in the potassium iodide solution forms silver iodide, etc. Cyanide of silver added to sodium sulphide changes into silver sulphide, as the cell: Silver, 0·1 normal silver nitrate solution, 0·1 normal sodium sulphide with silver sulphide, silver, has a higher electromotive force than the corresponding cyanide cell. On the other hand, silver sulphide does not dissolve in dilute potassium cyanide solution. The reason for this is easily seen when it is remembered that the more insoluble or complex a salt is, the lower is also the value of the product of the corresponding ions. If to a saturated silver chloride solution an amount of iodine ions (as in potassium iodide) be added equal to the chlorine ions present, silver iodide must precipitate; otherwise the product of the iodine and silver ions would be greater than is possible. This precipitation of silver iodide proceeds until the product of the ion concentrations has reached the constant value corresponding to the saturated silver iodide

N

solution. Such an arrangement of concentration cells is given in the following table of Ostwald (*Allgem. Chem.* ii. 882):—

	Volt.
Silver nitrate 0·1 normal—silver chloride in normal potassium chloride	0·51
Silver nitrate 0·1 normal—silver nitrate in normal ammonia	0·54
Silver nitrate 0·1 normal—silver bromide in normal potassium bromide	0·64
Silver nitrate 0·1 normal—silver nitrate in normal sodium thiosulphate	0·84
Silver nitrate 0·1 normal—silver iodide in normal potassium iodide	0·91
Silver nitrate 0·1 normal—silver nitrate in cyanide of potassium	1·31
Silver nitrate 0·1 normal—silver nitrate in normal sodium sulphide	1·36

Evidently the order of such a series may be changed by altering the concentrations of the electrolytes added to the silver salts. This might be done, for example, by adding a very concentrated solution of potassium chloride to the silver chloride solution; the concentration of the silver ions would thus be reduced below that of the 0·1 normal bromide solution, which contains silver bromide. In this case the electromotive force of the chloride cell would be greater than that of the bromide, and even if 0·1 normal potassium bromide solution be added to the chloride solution, silver bromide would not be precipitated; on the other hand, silver bromide could be dissolved in it. Similarly, silver sulphide would dissolve in concentrated potassium cyanide solution.

4. Finally, a concentration cell, which might also be included under description 1, may be here considered, because of its peculiar characteristics. Atten-

VI ELECTROMOTIVE FORCE 179

tion was first called to it by Ostwald. A cell consisting of one hydrogen electrode in an acid solution, and another in an alkali solution, the two solutions being in contact, is a concentration cell with regard to hydrogen ions. It has already been learned (p. 109) that water is slightly dissociated into H and OH ions, and consequently a certain quantity of H ions is present in the alkali solution. The electromotive force of this cell is

$$\pi = \frac{RT}{\epsilon_0} \ln \frac{p}{p_1},$$

p being the concentration or osmotic pressure of the hydrogen ions in the acid solution, and p_1 those in the alkali. Suppose the alkali and acid used to be normal solutions. The concentration (p) of the H ions in the acid solution, when the incomplete dissociation is taken into account, is about 0·8, and p_1 may be calculated from the measured electromotive force of the cell. In this case a considerable potential difference exists at the surface of contact between the two solutions, which must be taken into consideration, since the sum of the potentials at the electrodes alone is desired. With the correction given by Nernst,[1] the value of π at 18° is 0·81 volt; that is,

$$0\cdot81 = 0\cdot0575 \log \frac{p}{p_1},$$

or

$$\frac{p}{p_1} = 10^{14}.$$

Since p is 0·8, p_1, or the concentration of the hydrogen ions of the alkali solution, is $0\cdot8 \times 10^{-14}$.

According to the law of mass-action the product of

[1] *Zeitschr. physik. Chem.* xiv. 155, 1894.

the hydrogen and hydroxyl ions must again give a constant when divided by the concentration of the undissociated water. The latter concentration is so great as compared with the concentration of the ions, that it may be considered invariable. Consequently the product of the ion concentrations must also be practically constant. The concentration of the hydrogen ions of the alkali solution is 0.8×10^{-14}, as indicated above, and the concentration of the hydroxyl ions according to the supposition is 0.8; therefore the product is $(0.8)^2 \times 10^{-14}$. From this result the degree of dissociation of pure water may be directly ascertained. The product of the ion concentrations in pure water must be $(0.8)^2 \times 10^{-14}$, and the concentrations of the H and OH ions are here the same. If c be the concentration of one of these ions,

$$c^2 = (0.8)^2 \times 10^{-14} \text{ and } c = 0.8 \times 10^{-7}.$$

In other words, pure water is 0.8×10^{-7} normal with regard to its hydrogen or hydroxyl ions. The conductivity measurements of Kohlrausch (p. 111) gave 0.9×10^{-7}. This is a very remarkable agreement, and its significance is made greater by the fact that other methods for reaching the same end, as through the study of the hydrolysis of salts and the saponifying effect of water, have led to very nearly the same value.

Oxygen electrodes may be used instead of hydrogen, and the cell still have the same electromotive force, because the concentrations of the hydrogen ions in the two solutions are in the same relation to each other as those of the corresponding hydroxyl ions. This follows from the fact that the product of the concentrations of the H and OH ions of the solutions in the element is a constant. The fact that the platinum does not absorb oxygen as readily as it does hydrogen, and that it reaches a condition of equilibrium with the surrounding gas more slowly, makes it more difficult to obtain constant values. The current passes through the cell from the more concentrated solution to the other, as is always the case when the electrode produces negative ions. The direction of the current is considered as that in which the positive ions move.

It may be repeated here that, except for the potential difference existing between the solutions at their point of contact, the electromotive force of such cells does not depend upon the nature of the negative ion of the acid, nor upon the positive of the alkali. On the other hand, when acids of the same molecular concentrations are used, the degree of dissociation comes into play. The cell, hydrogen, acetic acid, potassium hydrate, hydrogen, would exhibit a lower electromotive force than the cell of corresponding concentration of hydrogen, hydrochloric acid, potassium hydrate, hydrogen. The slightly dissociated acetic acid contains less hydrogen ions than the highly dissociated hydrochloric acid; consequently in the latter cell the difference in concentration between the hydrogen ions of the acid and alkali solutions is greater than in the former, and therefore its electromotive force is also greater. That the same considerations apply to bases may be safely concluded from the measurements which have already been made in that direction.

C. Concentration Double Cells

Another form of concentration cell, differing from the previously described liquid concentration cell, is made by connecting two such cells as a double element. The so-called calomel cell, which is often used, serves as a type of this form. Its arrangement is as follows: Zinc, zinc chloride solution (concentrated), mercurous chloride, mercury, mercurous chloride, zinc chloride solution (dilute), zinc. The mercurous chloride is in excess, and covers the mercury. This cell differs from the simple cell: Zinc, zinc chloride solution (concentrated), zinc chloride solution (dilute), zinc, in having the combination mercurous chloride, mercury, mercurous chloride, between its two differently concentrated solutions of zinc chloride. Consequently the processes of electrolysis and the electromotive forces of these arrangements differ from those of the simpler cells. In the case, zinc, dilute zinc chloride solution, concentrated zinc chloride, zinc, when 2×96540 coulombs pass, there is a migration of zinc and chlorine ions from one solution to the other, and simultaneous solution and precipitation of two equivalents of zinc at the electrodes. In the calomel concentration cell such a migration cannot occur. When 2×96540 coulombs pass through this cell, two equivalents of zinc dissolve in the dilute chloride solution, and two of mercury separate from the mercurous chloride. Here the current always passes from the dilute to the concentrated solution within the cell. The mercury ions come from the dissolved mercurous chloride, and those precipitated are immediately replaced by the further solution of mercurous chloride. In the concentrated solution, on the other hand, two equivalents of zinc separate at the electrode, and two of mercury are dissolved. It must be borne in mind that when two equivalents of metallic mercury have been produced from the solid mercurous chloride in the dilute solution, two equivalents of chlorine ions have also been formed, and when two equivalents of metallic mercury have changed to mercurous chloride in the concentrated solution at the same time, two of chlorine ions have disappeared. When the quantities of the solutions are imagined so great that these changes take place without sensible influence on the concentration, the processes may be summarised as follows. Two equivalents of zinc and two of chlorine—that is, one gram molecule of zinc choride—have been transported from the concentrated solution to the dilute, while the quantity of mercury and of mercurous chloride remains unaltered. If the osmotic pressure of the zinc ions in the concentrated solution be p, and in the dilute solution p_1, then the corresponding osmotic pressures of the chlorine ions are $2p$ and $2p_1$. The maximum osmotic work is easily calculated, and is

$$\mathrm{RT} \ln \frac{p}{p_1} + 2\mathrm{RT} \ln \frac{2p}{2p_1} = 3\mathrm{RT} \ln \frac{p}{p_1}.$$

The electrical energy is $2\pi\epsilon_0$; therefore

$$\pi = \frac{3}{2} \frac{\mathrm{RT}}{\epsilon_0} \ln \frac{p}{p_1}.$$

In general

$$\pi = \frac{n_i}{n_e} \frac{\mathrm{RT}}{\epsilon_0} \ln \frac{p}{p_1},$$

where n_i is the number of ions produced by one

molecule of the electrolyte, and n_e the number of ϵ_0 necessary for the passage of a gram-molecule of the electrolyte from the concentrated to the dilute solution (see p. 169).

From the formula it may be seen that only the ratio $\dfrac{p}{p_1}$, n_t, and n_e have influence on the value of π. As Ostwald predicted, and as Goodwin [1] experimentally demonstrated, it follows that :—

1. The mercurous chloride and mercury of the calomel cell may be replaced by silver chloride and silver without altering the electromotive force.

2. Instead of zinc chloride, zinc bromide or iodide may be used when the depolariser [2] is a difficultly soluble bromide or iodide, without changing the electromotive force.

3. The electromotive force of the cell will not be changed if cadmium chloride and cadmium be substituted for zinc chloride and zinc.

4. If the zinc and zinc chloride be replaced by thallium and thallium chloride, the electromotive force will be considerably increased.

5. If instead of the chloride of zinc, the sulphate be used, with a difficultly soluble sulphate as depolariser, the electromotive force will be less than before. Whether lead or mercurous sulphate be used as depolariser can make no difference. The accompanying tables confirm these statements. For brevity the cells are designated by their soluble salts and depolarisers.

---

[1] *Zeitschr. physik. Chem.* xiii. 577, 1894.

[2] The difficultly soluble salt is here called a depolariser, because, through its presence, the electrode is made unpolarisable.

I

$ZnCl_2$ – HgCl and $ZnCl_2$ – AgCl Cells at 25°

Concentration of the $ZnCl_2$.	Observed E.M.F. of $ZnCl_2$ – HgCl.	Observed E.M.F. of $ZnCl_2$ – AgCl.	Calculated E.M.F. in Volts.
0·2 – 0·02	0·0787	0·0767	0·0797
0·1 – 0·01	0·0800	0·0780	0·0818
0·02 – 0·002	0·0843	0·0843	0·0844
0·01 – 0·001	0·0861	0·0847	0·0853

Considering the experimental errors of $1-2$ thousandths of a volt, the agreement is very satisfactory.

II

$ZnBr_2$ – HgBr and $ZnBr_2$ – AgBr Cells

Concentration of the $ZnBr_2$.	Observed E.M.F. of $ZnBr_2$ – HgBr.	Observed E.M.F. of $ZnBr_2$ – AgBr.	Calculated E.M.F. in Volts.
0·2 – 0·02	0·0793	0·0793	0·0797
0·1 – 0·01	0·0808	0·0802	0·0818
0·02 – 0·002	0·0860	0·0852	0·0844
0·01 – 0·001	0·0863	0·0858	0·0853

Through replacement of zinc and its chloride by cadmium and cadmium chloride, the value of the electromotive force could not be calculated, the concentration of the cadmium ions not being determinable (by the conductivity method). This is explained by the fact that $CdCl_2$ dissociates not only into Cd^{II} and Cl, Cl, but probably also, in concentrated solutions, into CdCl and Cl. In dilute solutions, where only the former dissociation is considerable, the values calculated agreed with those experimentally found.

III

TlCl – HgCl Cells.

Concentration of the TlCl.	Observed E.M.F.	Calculated E.M.F.
0·0161 – 0·00161	0·102	0·114
0·008 – 0·0008	0·100	0·115
0·0161 – 0·008	0·0328	0·033

IV

$ZnSO_4$ – $PbSO_4$ Cells

Concentration of the $ZnSO_4$.	Observed E.M.F.	Calculated E.M.F.
0·2 – 0·02	0·0427	0·0453
0·1 – 0·01	0·0440	0·0471
0·02 – 0·002	0·0522	0·0500

V

$ZnSO_4$ – Hg_2SO_4 Cells

Concentration of the $ZnSO_4$.	Observed E.M.F.	Calculated E.M.F.
0·2 – 0·02	0·047 – 0·034	0·045
0·1 – 0·01	0·045 – 0·033	0·047

The formula

$$\pi = \frac{n_i}{n_e} \frac{RT}{\epsilon_0} \ln \frac{p}{p_1}$$

is only applicable when the solubility of the depolariser is inappreciable. If, for example, the difficultly soluble mercurous chloride of the calomel cell be replaced by the comparatively easily soluble thallium chloride, it must be taken into account that the concentrations of the zinc and the chlorine ions are no longer in the same relation. Chlorine ions from the thallium chloride are thus added to those of the zinc chloride, and from the law of mass-action the product of the ion concentrations of the thallium and chlorine in the saturated thallium chloride solution is constant, and more chlorine ions must enter the dilute than the concentrated zinc chloride solution. From this consideration, taking into account the previous deduction, p and p_1 being the osmotic pressures or the concentrations of the zinc ions, and p' and p_1' those of the chlorine ions,

$$2\epsilon_0\pi = \mathrm{RT}\ln\frac{p}{p_1} + 2\mathrm{RT}\ln\frac{p'}{p_1'}$$

$$\pi = \frac{\mathrm{RT}}{\epsilon_0}\left(\frac{1}{2}\ln\frac{p}{p_1} + \ln\frac{p'}{p_1'}\right) \qquad (1).$$

In general

$$n_e\epsilon_0\pi = n_i\mathrm{RT}\ln\frac{p}{p_1} + n_i'\mathrm{RT}\ln\frac{p'}{p_1'},$$

where n_i and n_i' represent the number of cathions and anions which the molecule of the electrolyte produces, and n_e the number of ϵ_0 corresponding to the transportation of one molecule of the electrolyte from the concentrated to the dilute solution.

The electromotive force of the cell may also be calculated from the solution pressures of the two metals coming into consideration (in the calomel cell, the zinc and mercury). In this case the electromotive force of the cell consists of four potential differences, existing at the four points of contact between metal and liquid. If P_{Zn} and P_{Hg} represent the solution pressures of the zinc and mercury respectively, and p, p_1, p', and p_1' the concentrations of the zinc and mercury ions in the concentrated and in the dilute solutions, while n_{Zn} and n_{Hg} are the valencies of the metals, the electromotive force is represented by the following formula (the fact that the current passes in the element from the dilute to the concentrated solution being taken into account):

$$\pi = \frac{RT}{\epsilon_0}\left(\frac{1}{n_{Zn}}\ln\frac{P_{Zn}}{p_1} + \frac{1}{n_{Hg}}\ln\frac{p_1'}{P_{Hg}}\right.$$
$$\left. + \frac{1}{n_{Hg}}\ln\frac{P_{Hg}}{p'} + \frac{1}{n_{Zn}}\ln\frac{p}{P_{Zn}}\right).$$

This may be shortened to the form

$$\pi = \frac{RT}{\epsilon_0}\left(\frac{1}{n_{Zn}}\log\frac{p}{p_1} + \frac{1}{n_{Hg}}\log\frac{p_1'}{p'}\right),$$

or

$$\pi = \frac{RT}{\epsilon_0}\left(\frac{1}{2}\ln\frac{p}{p_1} + \ln\frac{p_1'}{p'}\right) \qquad (2).$$

Formulæ (1) and (2) lead to the same result, in spite of their apparent difference. In (1) $\dfrac{p'}{p_1}$ represents the concentration relation of all the negative ions of the solutions, while in (2) $\dfrac{p_1'}{p'}$ represents that of the cathions of the depolariser. It must be remembered that saturated solutions of the depolariser are being considered; consequently the product of the concentrations of all the anions and cathions of the depolariser is a constant (the anions of the electrolyte and depolariser being alike, as in the case of $ZnCl_2$ and $HgCl$).

The separate concentrations are also in a definite relation to each other. When, for instance, the cathions and anions are of the same valency, as in the example, their different concentrations in the solutions are inversely proportional to each other. If the anion be bivalent and the cathion univalent, the concentration of the latter is inversely proportional to the square of that of the former, and so on. This explains the agreement of the two formulæ.

LIQUID CELLS

It has already been stated in the considerations of the concentration cells that potential differences occur at the points of contact between the solutions. This assumption has been entertained a long time, but a clear conception of the derivation of such potentials did not exist. The Becquerel acid-alkali cell is well known; two platinum electrodes connected together are placed one into acid and the other into alkali solutions. That in the acid becomes positively and the other negatively charged; the potential difference varying with the conditions often amounts to more than 0·6 volt. Formerly the source of this electrical energy was erroneously thought to be in the heat generated by the neutralisation of the acid and alkali. As previously explained, this is practically a concentration cell. Oxygen of the air is present at the two electrodes, and in the acid solution there are few, while in the alkali there are many OH ions. Since the electrodes are of ordinary platinum instead of being coated with platinum black, it is easily explicable that the electromotive force of such a cell is variable. Ordinary platinum does not absorb oxygen to a very great extent, so that the condition of equilibrium which should be established, in which the concentration of the oxygen dissolved in the platinum corresponds to the pressure of the surrounding oxygen, as in the case of platinised platinum, is practically unrealisable; consequently the cell has an uncertain and varying value. This cell cannot generate a perceptible current, because the quantity of oxygen absorbed by the electrodes is very small, and, being exhausted, is replaced by that of the air only very slowly. The presence of other gases also has an influence upon the electromotive force of this cell.

We are indebted to Nernst[1] for satisfactory explanations of the phenomena of these liquid cells, their theory having been developed by him. If a solution of hydrochloric acid, for example, be placed in contact with a more dilute solution or with pure water, the acid will diffuse into the water. The hydrogen and chlorine ions of the acid are, to a certain extent, independent particles capable of moving with different velocities from places of higher osmotic pressure to those of lower. Since the hydrogen ions migrate more rapidly than those of chlorine, the foremost of the diffusing ions are hydrogen, and since these possess positive charges, the water or the dilute solution as a whole exhibits a positive, and the stronger solution a negative charge. Owing to the mutual attraction of the positive and negative charges of the hydrogen and chlorine ions, this separating process does not actually take place to any measurable extent, the hydrogen ions are delayed, and the chlorine ions increase their speed, so that a condition is reached in which both migrate at the same rate. The electrostatic

[1] *Zeitschr. physik. Chem.* iv. 129, 1889.

attraction, as well as the potential difference between the solutions, exists until both solutions are homogeneous. *The unequal velocities of migration of the ions are therefore the cause of the potential differences at the contact surfaces of differently concentrated solutions.*

If the negative ions have the greater velocity of migration, the more dilute solution will evidently be negative to the concentrated. In other words, *the dilute solution always presents the electricity of the more rapidly moving ion.*

Moreover, it is thus not only possible to foresee the nature of the potential difference at the point of contact between two liquids, but also in many cases quantitatively to calculate the magnitude of such potential differences, and prove the calculations by actual experiment. To illustrate this point, two differently concentrated solutions of an electrolyte, consisting of two univalent ions, may be imagined in contact. Let n be the share of transport of the positive ion, and $(1-n)$ consequently that of the negative. The quantity of electricity ϵ_0 is now conducted through the solutions from the concentrated to the dilute, then n positive gram-ions pass from the concentrated into the dilute, and at the same time $(1-n)$ negative gram-ions from the dilute into the concentrated solution. Let p represent the concentration of the positive and negative ions in the concentrated solution, and p_1 the same in the dilute solution. The maximum work, the process being completed osmotically, is

$$A = n\mathrm{RT}\ln\frac{p}{p_1} - (1-n)\mathrm{RT}\ln\frac{p}{p_1}$$

$$= (2n-1)\mathrm{RT}\ln\frac{p}{p_1},$$

or if n be replaced by $\dfrac{u}{u+v}$, u being the velocity of migration of the positive, and v that of the negative ions,

$$A = \frac{u-v}{u+v} \, RT \ln \frac{p}{p_1}.$$

Consequently

$$\pi = \frac{u-v}{u+v} \frac{RT}{\epsilon_0} \ln \frac{p}{p_1} \qquad (a)$$

because $\pi \epsilon_0 = A$.

If u be greater than v, the electric current passes from the concentrated to the dilute solution in the element itself; if v be greater than u, the current is in the opposite direction. If, finally, $u = v$, no potential difference exists between the solutions, and consequently there is no current.

Nernst arranged such liquid cells so that the only potential observed was that at the point of contact of two solutions, and compared the experimentally determined values of the electromotive force with those calculated from the formula. The following arrangement was used : Mercury—mercurous chloride
$$\overset{\text{I.}}{—}\text{0·1 normal KCl}\overset{\text{II.}}{—}\text{0·01 normal KCl—0·01 normal}$$
$$\overset{\text{III.}}{\text{HCl}}\overset{\text{IV.}}{—}\text{0·1 normal HCl—0·1 normal KCl—mercurous}$$
chloride—mercury. Since the two ends are identical, the potential differences occurring there neutralise each other, and only those differences at the four contact points I. II. III. and IV. are to be taken into account.

It is to be observed that, as far as experience has gone, the rule holds for liquid cells that only the ratio, not the absolute values of the osmotic pressures, comes into consideration. Therefore the potential difference of II. is equal and oppositely directed to that of IV.; thus the potential differences at I. and III. alone remain, and may be calculated from the above formula. If u_1 and v_1 are the velocities of migration of the potassium and chlorine ions respectively, while u_2 and v_2 ($= v_1$ here, because they represent the same negative ions) are the migration rates of the hydrogen and chlorine ions, then the sum of the potential difference is represented by

$$\pi = \frac{u_1 - v_1}{u_1 + v_1} \frac{RT}{\epsilon_0} \ln \frac{p}{p_1} - \frac{u_2 - v_2}{u_2 + v_2} \frac{RT}{\epsilon_0} \ln \frac{p'}{p_1'},$$

and as

$$\frac{p}{p_1} = \frac{p'}{p_1'},$$

therefore

$$\pi = \left(\frac{u_1 - v_1}{u_1 + v_1} - \frac{u_2 - v_2}{u_2 + v_2} \right) \frac{RT}{\epsilon_0} \ln \frac{p}{p_1}.$$

p and p_1 are the osmotic pressures or concentrations of the potassium and chlorine ions in the concentrated and dilute potassium chloride solutions, p_1 and p_1' the corresponding values of the hydrogen and chlorine ions in the corresponding hydrochloric acid solutions. The actual measured potential difference was -0.0357 volt, taken negative, since the current in the cell flowed in the direction IV. to I., and has, in the calculation, been considered positive when it passed from the concentrated to the dilute potassium chloride solution. The potential difference resulting from calculation by the formula, taking into consideration the incomplete dissociation of the substances, differs from the above by about ten per cent.

The formula (a) only permits of calculation of the potential difference at the points of contact of two differently concentrated solutions of one and the same binary electrolyte. If it is desired to make it applicable to electrolytes whose ions have different valencies, it takes the form

$$\pi = \frac{\dfrac{u}{n} - \dfrac{v}{n'}}{u + v} \cdot \frac{RT}{\epsilon_0} \ln \frac{p}{p_1} \qquad (b),$$

n representing the valency of the positive and n' that of the negative ion.

If two different electrolytes are in contact, as, for instance, potassium chloride and hydrochloric acid, the calculation is more difficult. Only for the case in which the total concentration of ions in each of the two solutions is the same, the following simple expression holds:

$$\pi = \frac{RT}{\epsilon_0} \ln \frac{u' + v''}{u'' + v'} \qquad (c),$$

where u' and v' are the migration rates of the ions of one electrolyte, u'' and v'' those of the other.

The calculation is still more difficult when one of the electrolytes contains polyvalent ions. If all the ions of the two solutions of binary electrolytes are polyvalent and of the same valency, then when the ion concentrations are the same,

$$\pi = \frac{RT}{n\epsilon_0} \ln \frac{u' + v''}{u'' + v'} \qquad (d).$$

It is worthy of special attention that in general there can be no arrangement of solutions in an electromotive series such as Volta formed for the metals. This is evident from the fact already mentioned, that such solution cells as the one measured by Nernst (p. 192) produce a current. A circuit consisting of metals only, at a common temperature, does not

generate an electric current.   If, on the other hand,
the solutions of the above cell, without the mercury
and mercurous chloride, be arranged
in a circuit, as shown in Fig. 24,
an electric current is obtained whose
electromotive force is that previously
calculated.   The existence of this
current may be demonstrated by
its power of induction, and it lasts
until the concentration of the
various ions is the same throughout the system.

FIG. 24.

The law of electromotive series applies only to
differently concentrated solutions of the same electrolyte in juxtaposition.   That it holds in this case may
be shown by adding the potential differences occurring
at the different points of contact, and comparing the
sum with the potential difference actually observed
between the first and last solutions placed directly in
contact.   The intermediate members of the series are
thus shown to play no part.

In considering concentration cells, such conditions
were usually chosen that the potential differences
occurring at the contact points of the solutions were
negligible.   Under such circumstances the electromotive force as previously given, for a cell in which
the metal electrodes dip into the two differently concentrated solutions of the salt, is

$$\pi = \frac{RT}{n_0 \epsilon_0} \ln \frac{p}{p_1}.$$

This formula was obtained by adding the potential
differences existing at the electrodes—that is, with the
application of the idea of electrolytic solution pressures.
In the addition the solution pressures were cancelled

from the equation, as they have the same value for the two similar electrodes and are oppositely directed.

It was also found possible to obtain the value of π, without any assumption of solution pressure, by the so-called purely energetic method. It was only necessary to take into account the condition of the system before and after the passage of a certain quantity of electricity, without attempting to understand why a potential difference and electric current are manifested. The maximum work obtainable osmotically by the change of the system from its original to its ultimate state is calculated, and this maximum is considered as the equivalent of the electrical energy. The values of π calculated in both ways agreed without exception.

It remains to be seen whether, when a potential difference occurs at the point of contact of the liquids, the two methods of calculation still yield the same result. For this purpose the concentration cell: zinc, zinc chloride (concentrated solution), zinc chloride (dilute solution), zinc is selected.

1. Calculation of π by means of the electrolytic solution pressure.

The electromotive force of the cell consists of three potential differences: the two at the electrodes and that at the point of contact between the two liquids. The sum of the first two is

$$\pi_{1\cdot 2} = \frac{RT}{2\epsilon_0} \ln \frac{p}{p_1},$$

where p and p_1 are the osmotic pressures of the zinc ions in the concentrated and dilute solutions respectively, the corresponding pressures of the chlorine ions being $2p$ and $2p_1$.

The third potential difference is calculated according to the formula (b), p. 194, and is

$$\pi_3 = \frac{\dfrac{u}{2} - \dfrac{v}{1}}{u+v} \frac{RT}{\epsilon_0} \ln \frac{p}{p_1},$$

where u and v are the rates of migration of the zinc and chlorine ions. The sum of $\pi_{1\cdot2}$ and π_3 is

$$\pi = \frac{RT}{\epsilon_0} \ln \frac{p}{p_1} \left(\frac{1}{2} - \frac{u-2v}{2(u+v)} \right) = \frac{3v}{2(u+v)} \frac{RT}{\epsilon_0} \ln \frac{p}{p_1},$$

or if the transportation ratios are introduced, $n = \dfrac{u}{u+v}$ and $1 - n = \dfrac{v}{u+v}$:

$$\pi = \frac{3(1-n)}{2\epsilon_0} RT \ln \frac{p}{p_1}.$$

π_3 must be subtracted from $\pi_{1\cdot2}$, as indicated, since the calculation of π_3 presupposes the direction of the positive current from the concentrated to the dilute solution within the cell, while with $\pi_{1\cdot2}$ the current passes in the opposite direction.

2. Calculating with respect to the energy change alone, the process is exactly that outlined on p. 192. If $2\epsilon_0$ be allowed to pass through the cell, a gram-ion of zinc passes into the dilute, while the same quantity is deposited from the concentrated solution. In addition, the quantity n gram-ions of zinc pass from the dilute to the concentrated solution, n being the transportation share of the zinc ions. The dilute solution is now richer by $(1 - n)$ gram-ions of zinc, while the concentrated one has lost this amount. Simultaneously, however, an amount of chlorine ions equivalent to the $(1 - n)$ zinc ions has also passed from the concentrated to the dilute solution; consequently the

quantity $(1 - n)$ of zinc and its equivalent of chlorine ions have been moved from the concentrated to the dilute solution. The maximum osmotic work corresponding to the zinc ions is

$$(1 - n)\mathrm{RT} \ln \frac{p}{p_1},$$

and since there are two chlorine ions to each zinc, it has for the chlorine ions the value

$$2(1 - n)\mathrm{RT} \ln \frac{p}{p_1},$$

or, added together,

$$3(1 - n)\mathrm{RT} \ln \frac{p}{p_1}.$$

The electrical energy is $2\epsilon_0\pi$, and therefore

$$\pi = \frac{3(1 - n)}{2\epsilon_0} \mathrm{RT} \ln \frac{p}{p_1},$$

the same as the previous formula.

This agreement in the methods gives also a method for determining the magnitude of potential differences at the contact points of liquids. It is only necessary to calculate as above, the sum of the potential differences occurring at the two electrodes, and subtract it from the actually measured electromotive force of the whole cell, to obtain the desired value.

GENERAL CONSIDERATION OF CONCEN- TRATION AND LIQUID CELLS

All the cells thus far described have the common characteristic that *their electrical energy is not generated from chemical energy.* In every case there was simply a passage of material from a higher to a lower pressure, and whether it be a gas or a dissolved substance which undergoes this change, the process does not affect the internal energy. The work done does not therefore come from the internal energy, but is derived from the heat of the surroundings. Consequently *the galvanic cells thus far considered are really machines for turning the heat of their surroundings into electrical energy.*

According to the generally applicable formula of Helmholtz (see p. 142),

$$\epsilon_0 \pi - Q = \epsilon_0 T \frac{d\pi}{dT}.$$

In the present case Q, the heat generated by the chemical reaction, is zero ; therefore

$$\epsilon_0 \pi = \epsilon_0 T \frac{d\pi}{dT} ; \text{ or } \frac{\pi}{T} = \frac{d\pi}{dT} ; \text{ and } \pi = T \frac{d\pi}{dT}.$$

This, on integration, gives $\ln \pi = \ln T + k$ or $\frac{\pi}{T} = k$.

The change of the electromotive force of these cells with the temperature is determined by the relation existing between the electromotive force and the corresponding absolute temperature. The electromotive force itself is proportional to the absolute temperature. When in activity the cell cools itself and takes up heat from the surroundings.

The same conclusions are reached on proceeding in still another way. The electromotive force of one of the previously mentioned concentration or liquid cells is in general

$$\pi = x \frac{RT}{\epsilon_0} \ln \frac{p}{p_1} \qquad (a),$$

from which

$$\frac{\pi}{T} = x \frac{R}{\epsilon_0} \ln \frac{p}{p_1} \qquad (b).$$

On differentiation with respect to T,

$$\frac{d\pi}{dT} = x\frac{R}{\epsilon_0}\ln\frac{p}{p_1}$$ (c)

is obtained, if x and $\ln\frac{p}{p_1}$ for " ideal " solutions are considered as practically independent of the temperature.

By combination of (b) and (c)

$$\frac{\pi}{T} = \frac{d\pi}{dT}$$

is again obtained.

It will be well to bear in mind that the electromotive force is only correctly calculable by this method
when the solutions are so dilute that the laws of gases
are applicable, for it is upon this assumption that the
maximum work is estimated. Moreover, it must be
possible to obtain the total energy in the form of
electricity. Since, as a matter of fact, neither of these
limitations is actually reached, the observed electromotive force cannot exactly agree with that calculated.
Regarding the first point the error is not negligible.
One proof of this is that solutions are often used
which, on being mixed, generate considerable quantities
of heat, and are therefore far from being ideal solutions.
For such solutions the Q of Helmholtz's formula is
evidently not zero, and the relation $\frac{\pi}{T} = \frac{d\pi}{dT}$ no longer
holds good.

From these observations it is furthermore evident
that it is unreasonable to consider the heat generated
by the mixing of solutions used in the cells as the
source of, or the reason for, the electrical energy produced. In the concentration cell, for example,
platinum black with hydrogen, alkali, acid, plati-

num black with hydrogen, the electromotive force depends principally upon the difference of the concentration of the hydrogen ions in the two solutions. The process of neutralisation which may take place at the point of contact between the alkali and acid is not to be considered as determining the electromotive force of the cell, nor can it be looked upon as the principal reason for its existence. The same considerations apply to the cells in which an electrode is covered with one of its difficultly soluble salts, as, for example, mercury, mercurous chloride with potassium chloride, mercurous nitrate, mercury. The process of solution of the mercurous chloride has nothing directly to do with the production of the electromotive force.

THERMOELEMENTS—THE ELECTROMOTIVE SERIES

In connection with the foregoing a few words may well be devoted to the thermoelements. Heat is here subjected to a transformation into electrical energy caused by a difference of temperature. On the other hand, in the concentration cells heat at a constant temperature is changed into electricity. This cannot be considered as contrary to the second law of thermodynamics, because, according to this law, it is only in a *cyclic process* that no heat at constant temperature can be changed into work. In other processes such a transformation may well occur.

The potential difference at one electrode may be expressed by the formula,

$$\pi = \frac{RT}{n_e \epsilon_0} \ln \frac{P}{p},$$

and is accordingly proportional to the absolute temperature. The arrangement : zinc, zinc sulphate solution, zinc, will produce no electrical energy at constant temperature, since the two potential differences of such a cell are equal and oppositely directed. But if one of the contact points between electrode and solution be warmed, the corresponding potential difference changes and an electric current is produced. As the potential difference at the point of contact between two solutions is also proportional to the absolute temperature, it is immediately clear that the following cyclic arrangement should produce an electric current :

       Solution of concentration C_1 at temperature T_1
·   Solution of concentration C_2 at temperature T_1
       Solution of concentration C_2 at temperature T_2
       Solution of concentration C_1 at temperature T_2

Since the osmotic pressure, the solution pressure, and the transportation ratios are functions of the temperature, the electromotive force of a thermoelement cannot be simply calculated. For further considerations of this point the reader is referred to the original work of Nernst (*Zeit. physik. Chem.* iv. 169, 1889).

More important to us than these thermoelements are those in which only conductors of the first class enter. In this case the measurement of the electromotive force, when the temperature difference between the points of contact is known, gives a method of determining the potential difference actually existing between two metals when at the same temperature. Since a thermoelement generates an electric current by the change of heat energy only into electricity, the equation of page 199 applies :

$$\frac{\pi}{T} = \frac{d\pi}{dT} \; ; \; \pi = T\,\frac{d\pi}{dT},$$

and this applies equally well to the combination as a whole as to the individual potential differences, since a cell can always be conceived in which there exists only the potential difference considered. It is, therefore, only necessary to know the change of the potential with the temperature $\left(\frac{d\pi}{dT}\right)$ at the point of contact between two metals, in order to be able to calculate π, or the potential difference at the temperature T. The value of $\frac{d\pi}{dT}$ may be directly obtained from the electromotive force of a thermoelement consisting of the two metals in question, the temperature at one contact point being T, and that at the other $T + dT$. If the temperature T is common throughout, the electromotive force is zero, as the two potential differences are equal and opposite. It is only because one of the potential differences may be changed by a temperature change that the electromotive force assumes a certain value, namely, that of the alteration in the potential difference. From the formula it is evident that if dT is unity, the electromotive force of the element is $Td\pi$.

The values of π, calculated for pairs consisting of the most widely differing metals at the ordinary temperature, are very small, and amount, even in exceptional cases, to but a few hundredths of a volt. In the preparation of thermopiles the latter metals or alloys are especially valuable. The above results are in perfect agreement with the previous assumption that in the majority of cells the principal source of the electrical energy is at the surface between electrode and solution.

The law of the electromotive series must evidently apply to the minute potential differences existing

between the metals themselves.   A cell composed of
only two metals cannot, therefore, generate an electric
current when the temperature is the same throughout.
This conclusion is necessitated by the second law of
thermodynamics, otherwise any desired quantity of
heat at constant temperature could be changed into
electrical energy without any permanent alteration
taking place in the system ; which is equivalent to
saying that a cyclic process may continually change
heat into work.   That this electromotive series exists
does not explain that discovered by Volta, since in
the latter the forces are very much greater.   Volta
thought that the potential difference now ascribed to
the surface between liquid and metal was really produced at the contact point between the metals.   To
corroborate his conclusions, the existence of a similar
law governing the potential differences at the surface
between metals and liquids must be demonstrated.

In the following pages it will be seen that, theoretically, a certain definite potential difference exists
between a metal and an electrolyte.   If, for example,
zinc, in contact with an electrolyte whose potential
is zero, exhibits a potential of 3, while cadmium is 2
and copper 1, then, according to the electromotive
series, the potential difference between zinc and copper
must be equal to the sum of that between zinc and
cadmium and that between cadmium and copper.   As
this is actually the case, the law of electromotive series
may be considered correct.   Very accurate measurements with an electrometer would evidently give slight
deviations, because in all cases another metal is
brought into contact with that of the electrometer, and
thus also another, though possibly very small, potential
difference is introduced.   In a similar manner the

electromotive series is roughly applicable to the galvanic cells. The arrangement : zinc, zinc sulphate, cadmium sulphate, cadmium, cadmium sulphate, copper sulphate, copper, in accordance with this law, should exhibit the same electromotive force as the combination : zinc, zinc sulphate, copper sulphate, copper, the concentrations of the zinc and copper sulphate solutions being the same in both cases. This is only exceptionally the case because of the disturbing influence of the potential differences at the surfaces between solutions. That the law applies to simple liquid cells in a certain definite case only has already been mentioned.

CHEMICAL CELLS

A distinction is to be made between the previously described cells, in which heat, and the "chemical cells," in which chemical energy is changed into electrical energy. A type of this latter class is the well-known Daniell element : zinc, zinc sulphate, copper sulphate, copper. When in activity zinc passes from the metallic into the ionic, and copper from the ionic into the metallic state. In this process (in contradistinction to the ideal concentration cells) a change in the internal energy of the system takes place, and this difference in energy may be considered as the principal source of the electrical energy produced. Instead of the change of positive ions to metal at one pole, and the metal to ions at the other, the negative ions may also perform this process. The cell, platinised platinum in oxygen gas, potassium hydrate, potassium chloride, platinised platinum in chlorine gas, causes OH ions to be produced in the alkali solution, and chlorine ions to change into molecular chlorine in the potassium chloride solution. (The current and process may be reversed under certain circumstances.)

Finally, positive ions may form at one electrode simultaneously with the negative ions at the other. An example is seen in the combination: zinc, zinc sulphate, potassium chloride, platinised platinum in chlorine gas. It is also well to remember that in all such cells there is a small potential difference produced at the surface between the solutions.

The electrical energy may be calculated by the Helmholtz formula, from the heat generated by the chemical processes and the experimentally determined temperature coefficients. The element during activity must yield as electrical energy the maximum work obtainable through the change of state. This work bears that relation to the heat of the chemical reaction measured in the calorimeter which is given by the Helmholtz formula. As this formula shows, there may be elements in which the chemical or internal energy change is exactly equal to the electrical energy obtained. These may be considered as machines which, in their action, will change all the energy put into them into another energy form. There are also cells in which only a portion of the chemical becomes electrical energy, and these may be looked upon as machines which transform only a portion of the energy introduced into another form of available energy, while the remainder is lost as heat. A third kind of cell is also known, by which more electrical energy is produced than corresponds to the chemical reactions taking place, and such elements may be considered as machines transforming not only the applied energy into work, but absorbing and changing into work the heat of the

surroundings.   Imagine in this last class the amount
of work which really comes from the heat of the
surroundings, continually increased ; cells are finally
reached in which (as in the concentration cells) the
internal energy remains unaltered and the electrical
energy is derived entirely from the heat of the surroundings.   It then becomes a question whether these
are to be designated chemical cells or not.

From these remarks it may be seen that a sharp
line of demarcation between the chemical and other
cells does not exist, and one form passes into the
other.   The distinction is justifiable in so far as the
chemical reaction is the chief characteristic of the
cells, the only other cells at constant temperature
being liquid and concentration cells, where there is no
chemical reaction.   Here, naturally, the Helmholtz
formula gives no aid, because of its general nature.

Again employing the idea of electrolytic solution
pressure, the electromotive force of the Daniell cell
may be represented by the formula (see p. 154)

$$\pi = \frac{RT}{2\epsilon_0} \ln \frac{P}{p} - \frac{RT}{2\epsilon_0} \ln \frac{P'}{p'} = \frac{RT}{2\epsilon_0} \left(\ln \frac{P}{p} - \ln \frac{P'}{p'} \right).$$

The inconsiderable potential difference between the
solutions is here omitted.

In writing the formula it was assumed that the
current passes from the zinc through the solution to
the copper.   If this were not the case, a negative
value would be obtained for π on taking the difference
between the separate potential differences, which would
signify that the current was oppositely directed.   It
is evidently impossible to foresee whether P is less
or greater than p, and whether P' is less or greater
than p', that is to say, whether the expressions for the

separate potential differences are positive or negative; but, as seen, it is unnecessary to give attention to this point. When it is desired to represent the electromotive force of a cell in which the electrodes yield only positive ions, as composed of the single potential differences, it is only necessary to represent the value for each electrode in the form

$$k \ln \frac{P}{p}$$

and write the expressions after one another. Having arbitrarily established the direction of the current, a positive sign is placed before the expression for that electrode which produces positive ions when the cell is active, and a minus sign before the expression corresponding to the electrode where positive ions leave the solution. The sum of these quantities is then the desired value.

For those cells or systems in which the current is due to changes in the negative ions alone, those expressions corresponding to electrodes at which negative ions disappear are to be written with the negative, and the others with the positive sign. For the cell: platinised platinum in oxygen gas — potassium hydrate solution — potassium chloride solution — platinised platinum in chlorine gas, it may be assumed that the current, that is, the positive electricity, passes from the oxygen electrode through the solutions to the chlorine electrode; then

$$\pi = \frac{RT}{\epsilon_0} \left(\ln \frac{P}{p} - \ln \frac{P'}{p'} \right),$$

where P and p are the solution pressure of the chlorine and osmotic pressure of the chlorine ions, P' and p' the solution pressure of the oxygen and osmotic pressure of the hydroxyl ions.

If, finally, both kinds of electrodes are present, special care must be exercised in order to avoid mistakes in the signs used. Those expressions are considered positive which represent electrodes where positive or negative ions are produced, and the minus sign is applied where positive or negative ions disappear. Accordingly, the electromotive force of the system : zinc — zinc sulphate solution — potassium chloride solution — platinised platinum in chlorine gas — platinised platinum in oxygen — potassium hydrate solution — copper sulphate solution — copper, the direction of the current being assumed to be from zinc to copper through the solutions, is

$$\pi = \frac{RT}{2\epsilon_0} \ln \frac{P}{p} + \frac{RT}{\epsilon_0} \ln \frac{P'}{p'} - \frac{RT}{\epsilon_0} \ln \frac{P''}{p''} - \frac{RT}{2\epsilon_0} \ln \frac{P'''}{p'''}$$

$$= \frac{RT}{2\epsilon_0} \left(\ln \frac{P}{p} - \ln \frac{P'''}{p'''} \right) + \frac{RT}{\epsilon_0} \left(\ln \frac{P'}{p'} - \ln \frac{P''}{p''} \right).$$

P, P', P'', P''' represent the solution pressures of the zinc, chlorine, oxygen, and copper respectively ; p, p', p'', and p''' the corresponding osmotic pressures of the ions.

In order thus to carry out the calculations of the electromotive forces, the solution pressures must be known. To learn these it is necessary to know some one potential difference (π) at the electrodes in question, from which, at known osmotic pressure, the required magnitude may be determined once for all, since all the values excepting P in the formula

$$\pi = k \ln \frac{P}{p}$$

are known.

DETERMINATION OF SINGLE POTENTIAL DIFFERENCES

By the experimental investigations of Lippmann upon the connection existing between the surface tension of mercury in sulphuric acid and the potential difference at the point of contact, the measurement of single potential differences was first made possible. The principal result of Lippmann's research was expressed by him as follows: The surface tension at the contact surface between mercury and dilute sulphuric acid is a continuous function of the electromotive force of the polarisation at that surface.

Helmholtz later made the researches of Lippmann better understood by an application of the theory of electrical double layers. If mercury be brought into contact with a liquid, *e.g.* dilute sulphuric acid, it assumes a positive electrical charge. This may be explained by assuming that the electrolyte contains mercury ions, very possibly from the dissolving of a little oxide, which may be present on the surface of even the purest mercury. The work of Warburg has also shown that the mercury may be oxidised by the oxygen dissolved in the liquid, and may thus enter the ionic state. Because of its very low solution pressure the mercury itself is positively charged in a solution containing very few of its ions.

On account of the electrostatic attraction, a number of negative ions group themselves about the positive electrode, and a double layer is formed (see also p. 148). If it be assumed that the mercury is "polarisable," *i.e.* no ions can pass from the mercury to the solution nor in the opposite direction (a condition only approximately attained), and if negative electricity be added to the mercury surface, a portion of the positive charge there present is removed, and at the same time the surface tension of the mercury is increased. This is the result of the mutual repulsion of the quantities of positive electricity on the surface of the mercury as well as the negative in the electrolyte, with the consequent expansion of the surface in opposition to the surface tension. If a portion of this electricity be removed, the surface tension naturally increases. By continued introduction of negative electricity a condition may be reached in which the double layer disappears and the surface is electrically neutral. Evidently at this point the surface tension has reached its maximum value. If still more negative electricity be introduced, the mercury becomes negatively charged, and the attracted positive ions of the solution form a new double layer, differing from the former in the relative position of the two kinds of electricity. The surface tension of the mercury must now decrease with increased negative charges at the surface because of the mutual repulsion of the quantities of electricity.

It is desired to ascertain the potential difference brought about by the electrostatic attraction of the double layer when the mercury in ordinary condition is immersed in dilute sulphuric acid. In order to make the mercury just neutral a potential difference must be brought about equal to that of the electrostatic attraction. Consequently that potential difference at which the maximum surface tension of the mercury is reached, when the latter is connected with the negative pole of a source of electricity, is the desired value. The mercury in this case does not differ in potential from the liquid, there being no double layer present.

The execution of the above experiment is simple in principle; the difficulties which have practically to be overcome in accurate investigations need not be discussed here. The apparatus depicted in Fig. 25 [1] may be used. The capillary c, as well as the greater part of the tube A, attached to c by a rubber tube, are filled with mercury. c dips into the cup B, which contains

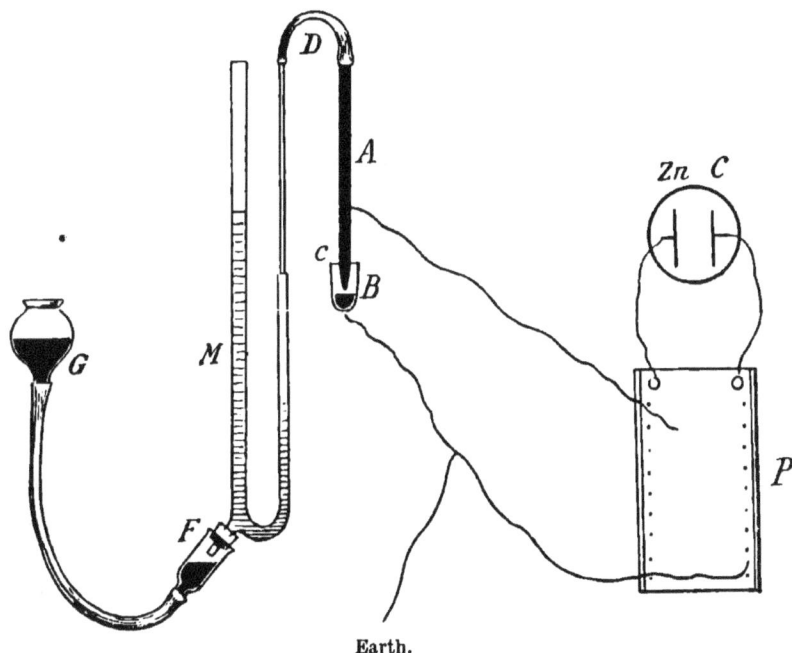

Fig. 25.

a little mercury, and above this the electrolyte. The position of the mercury in the capillary is observed by means of a microscope. The bulb G, which contains mercury, permits of the application of desired pressures through its elevation and depression; it is attached to the manometer (M) by a rubber tube. A bent glass tube D leads from the latter to A, the connections

[1] *Zeitschr. physik. Chem.* xv. 1, 1894.

being made with short pieces of rubber tubing. Paraffin oil serves as the liquid of the manometer, increasing the delicacy of the reading. A small vessel, as shown at F, containing both paraffin oil and mercury, is connected to the apparatus between the manometer and rubber tube. P is an arrangement for the introduction of any desired potential difference (see p. 125).

It is to be recalled that when a capillary is placed in water, the latter rises to a level above that of the surrounding liquid, as it wets the surface of the glass. On the other hand, with mercury the level in the capillary is below that of the surrounding liquid, and, if the surface tension be increased, sinks still lower, that is, it moves against the pressure of the mass of mercury. It is only in this way that a diminution of the surface, the result of increased surface tension, can occur.

If now a certain potential from the source of electricity be applied to the mercury in the capillary c, the surface tension of the mercury increases and the meniscus begins to rise. In order to hold this in its original position, a certain pressure must be exerted by means of the manometer. As the applied potential difference is increased the necessary pressure also increases, until at a certain potential difference a maximum in the pressure is observed, which, on further increase of the potential difference, again diminishes. The potential difference corresponding to the maximum pressure is that which is naturally assumed by the mercury in the electrolyte.

In order that the results may not be variable, it is necessary to add some mercury salt to the electrolyte, that this may have a certain concentration of mercury ions throughout, since the potential difference of the metallic mercury is dependent thereon. The question is naturally raised : Is not the electrode an unpolarisable one when sufficient mercury ions are present ? Why can it be considered as almost perfectly unpolarisable, as has been done ? In answer, attention is directed to the following : By adding mercury ions to the liquid, the mass of mercury in B becomes a nearly unpolarisable electrode, which maintains the same potential difference towards the electrolyte, no matter what other potential differences are inserted at P. Because of its small surface the metallic mass in the capillary only comes into direct contact with a very small part of the electrolyte. Consequently on the application of a potential difference only very few mercury ions pass from the electrolyte into metallic mercury, and new ions can diffuse into the layer at the surface but slowly ; therefore this electrode is practically polarisable. The relative extent of the surfaces of mercury evidently plays the important part. What is actually measured is the potential difference at the larger mercury surface, since this alone is constant. When the two quantities of mercury are in connection, that in the capillary changes its surface tension until it possesses the same potential difference as the lower mass. Such is evidently also the case when the larger electrode is an amalgam instead of pure mercury. For instance, if it be copper amalgam and the solution above it contains a copper salt, the potential difference will be less than before, since the amalgam assumes a less positive charge. The mercury in the capillary again assumes the potential of the lower electrode when the two are connected, and on introducing external potential differences a lower value than with pure mercury is sufficient to bring about the maximum surface tension.

There is a second method which can be used for the determination of single potential differences, the principle of which was explained by Helmholtz. Ostwald [1] first showed that it could be used for this purpose, and through his efforts, as well as those of Paschen, the method has been developed.

If an insulated mass of mercury be allowed to flow in a stream through a fine opening and drop into an electrolyte, there ought to be (in the ideal case) no potential difference between the mercury and the electrolyte. As already seen, mercury in contact with an electrolyte becomes charged with positive electricity. By allowing the mercury to drop into the electrolyte .the area of its surface is continually increased, and the charge must spread over the entire surface; in other words, the potential difference between mercury and electrolyte must approach zero. Helmholtz expressed himself on this point in the following manner :—

"Consequently I conclude that when a quantity of mercury is connected with an electrolyte by a rapidly dropping fine stream of the mercury, and is otherwise insulated, the two cannot possess different electrical potentials, for if a potential difference did exist, for example, if the mercury were positive, each falling drop would form an electrical double layer on its surface, requiring the removal of positive electricity from the mass, and diminishing its positive charge until that of the mercury and solution reached equality."

An experiment by A. König had already shown that the charge of the mercury could be almost com-

[1] *Zeitschr. physik. Chem.* i. 583, 1887.

pletely removed in the manner described. This result
was later confirmed in other ways. Fig. 26 represents
the arrangement employed by König.
The mercury cup (a), beneath dilute
sulphuric acid, was connected by a
wire (c), with mercury dropping
from the capillary into the acid.
A galvanometer (G) was connected into the circuit as shown.
This indicated that the positive
electricity was removed with the
dropping of the mercury in agreement with the previous explanations.

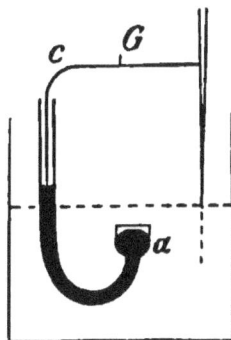

FIG. 26.

If the upper mercury, through the dropping, be
brought to practically the same potential as the
solution, the polarisable mercury in the cup has the
same potential, and therefore the maximum surface
tension. This could be determined by means of an
ophthalmometer. As still further proof, a weak
electromotive force, positive or negative, on being
introduced into the circuit on the wire connecting the
two electrodes, caused the surface tension to decrease,
since a potential difference was produced between the
liquid and the mercury of the cup. In this case it is
desirable to have the electrode as polarisable as possible,
for the potential of an unpolarisable dropping electrode
cannot be altered in this way.

By these two methods it is possible to ascertain the
magnitudes of the individual potential differences
which constitute the electromotive forces of reversible
cells. Obviating as far as possible the potential differences at the contact surface between the solutions by
suitable choice of electrolytes and concentrations, any
potential which an electrode assumes when in contact

with a liquid containing the corresponding ions (otherwise variable values are obtained) can be determined. On the one hand, the potential difference between mercury and a normal potassium chloride solution, for example, saturated with mercurous chloride, may be obtained by the pressure method. Then this electrode of known potential difference may be used in connection with the one whose potential difference is to be determined. Supposing the potential difference of silver in contact with a normal silver nitrate solution to be desired, the electromotive force of the combination : mercury — normal potassium chloride solution saturated with mercurous chloride — normal silver nitrate solution — silver, is measured. From this the value of the potential difference due to the mercury electrode is subtracted, and the desired potential difference remains. On the other hand, the same end may be reached by arranging the mercury in the above combination as a drop electrode, whereby the mercury and electrolyte are brought to the same potential. The measured electromotive force of the cell has its origin in the potential difference at the silver, and represents the latter. The mercury may be allowed to drop directly into the second electrolyte, which is often preferable.

Results obtained by these two methods agree satisfactorily, although differences of a few hundredths of a volt exist, probably due to the difficulty of measurement. For the determination of single potential differences it is customary to make use of a so-called "normal" electrode [1] as shown in Fig. 27. At the bottom of a small upright vessel, about 8 cm. high and 2 or 3 cm. in diameter, pure mercury is

---

[1] Ostwald, *Phisiko-chemische Messungen*, p. 258.

placed, and is covered with a layer of mercurous
chloride; the vessel is then filled with a normal
potassium chloride solution and closed with a rubber
stopper carrying two glass tubes.   Through one of
these a platinum wire is introduced, which is in contact with the mercury; the other tube is filled with
the chloride solution, which also fills the rubber tubing
and bent glass tube at its end.   The latter is placed
into the liquid, the potential difference between which

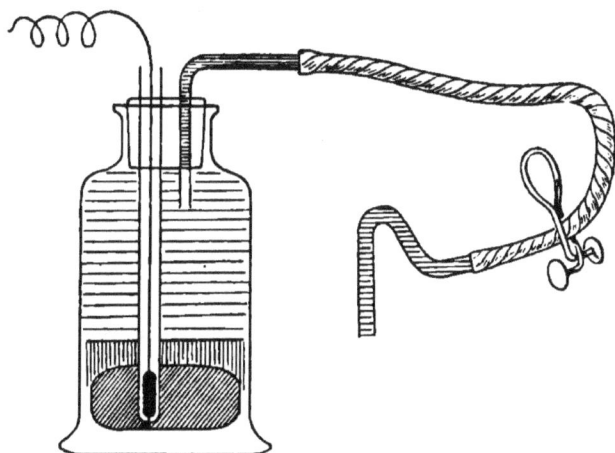

FIG. 27.

and an electrode is to be determined, and the electromotive force of the cell thus formed is measured.
If the potassium chloride produces a precipitate
with the second electrolyte, as with silver nitrate, a
third and indifferent solution, e.g. sodium nitrate, is
introduced between the two.   The use of potassium
chloride solution for the normal electrode offers the
advantage that it does not favour the formation of
potential differences at the contact points of the
solutions, since the rates of migration of its ions

are very nearly the same. The potential difference produced between the solutions at their point of contact is a disturbing factor, and affects the value of single potential differences by several thousands of a volt. It may even amount to a few hundredths under certain circumstances.

At present 0·56 is accepted as the most probable value for the potential difference between mercury and normal potassium chloride solution saturated with mercurous chloride. The mercury ions possess the tendency represented by 0·56 volt of leaving the ionic and entering the metallic state. The metal is therefore positively and the electrolyte negatively electrified. This fact must always be borne in mind in order to be able properly to carry out the calculation. Hereafter the potential of the metal or electrode will be considered as zero and the + or − sign will indicate whether the electrolyte is positive or negative to the electrode.[1] In accordance therewith mercury − mercurous chloride in normal potassium chloride solution = − 0·56 volt.

Through the aid of this value, any other single potential difference may be determined. Suppose the electromotive force of the cell: zinc − normal zinc sulphate solution − mercurous chloride in normal potassium chloride solution − mercury, has been measured and found to be 1·08 volt, and that the current passes from the zinc to the mercury through the solutions, the potential difference, zinc − normal

---

[1] Another form of expression occurs in the literature. The + sign is used by some investigators to designate that potential difference by the action of which ions are produced, no matter whether these be positive or negative. The above use of the signs is considered simpler for the calculations.

zinc sulphate, may be calculated as follows. According to p. 208 the electromotive force of the cell is

$$\pi = \frac{RT}{2\epsilon_0} \ln \frac{P}{p} - \frac{RT}{\epsilon_0} \ln \frac{P'}{p'} = \pi_1 - \pi_2,$$

P and p referring to the zinc, P' and p' to the mercury.

$$\pi = 1{\cdot}08,$$

and also

$$\pi_2 = -0{\cdot}56 \; ;$$

therefore

$$1{\cdot}08 = \pi_1 + 0{\cdot}56,$$

and

Zinc – normal zinc sulphate $= \pi_1 = +0{\cdot}52$ volt.

Zinc has the tendency, represented by the electromotive force 0·52 volt, to send its ions into the normal solution of its sulphate, and the solution is therefore positively while the metal is negatively electrified.

Experience has shown that this calculation is not at first easily understood by students, and therefore two other illustrations are given.

A measurement of the cell : copper – normal copper sulphate solution – mercurous chloride in normal potassium chloride solution – mercury, gave 0·025 volt as electromotive force, the direction of the positive current being from the mercury to the copper through the electrolytes.

$$\pi = -\frac{RT}{2\epsilon_0} \ln \frac{P}{p} + \frac{RT}{\epsilon_0} \ln \frac{P'}{p'} = -\pi_1 + \pi_2,$$

P and p referring to the copper, and P' and p' to the zinc. Since $\pi = 0{\cdot}025$ volt and

$$\pi_2 = -0{\cdot}56 \text{ volt,}$$

therefore

$$0\cdot025 = -\pi_1 - 0\cdot56$$

and

Copper – normal copper sulphate solution $= \pi_1 = -0\cdot585$ volt.

The cell, consisting of platinised platinum in oxygen at atmospheric pressure – normal sulphuric acid – mercurous chloride in normal potassium chloride solution – mercury, exhibited an electromotive force of $0\cdot75$ volt, the current passing from the mercury to the oxygen through the electrolytes. Since

$$\pi = +\frac{RT}{\epsilon_0}\ln\frac{P}{p} + \frac{RT}{\epsilon_0}\ln\frac{P'}{p'} = -\pi_1 + \pi_2,$$

where P and p refer to oxygen and P$'$ and p' to the mercury, therefore

$$0\cdot75 = -\pi_1 - 0\cdot56.$$

Hence it is seen that the combination : oxygen under atmospheric pressure – normal sulphuric acid $= \pi_1 = -1\cdot31$ volt.

Oxygen has therefore the tendency to generate OH ions with a considerable electromotive force. The electrolyte thereby becomes negatively and the electrode positively charged.

In the avoidance of error the following method of consideration is particularly advantageous. The electromotive force and direction of current of the measured cell and of the normal electrode are known, also the fact that the total electromotive force is composed of the two single values ; therefore for every cell the following graphic scheme may be adopted, and is here applied to the second of those above described :

Copper $- \, ^{1}/_{1}$ n. copper sulphate $-$ mercurous chloride, etc. $-$
mercury.

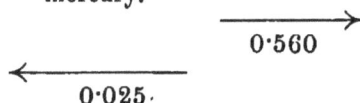

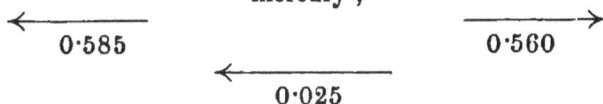

$$\xrightarrow{\hspace{3cm}} \\ 0\text{·}560$$

$$\xleftarrow{\hspace{2.5cm}} \\ 0\text{·}025\,\cdot$$

The third value is now definitely determined. If
there is a potential difference between the mercury and
its electrolyte of 0·560 volt, and its direction be represented by the arrow, while the electromotive force
of the combination is 0·025 volt in the opposite
direction, it is evident that between the copper and
its electrolyte there must be a potential difference of
0·585 volt in the same direction as the total electromotive force of the cell, or

Copper $-$ n. copper sulphate $-$ mercurous chloride, etc. $-$
mercury ;

$$\xleftarrow{\hspace{2.5cm}} \qquad\qquad \xrightarrow{\hspace{2.5cm}} \\ 0\text{·}585 \qquad\qquad\qquad\qquad 0\text{·}560$$

$$\xleftarrow{\hspace{2.5cm}} \\ 0\text{·}025$$

and

Copper $-$ n. copper sulphate solution $= \, -0\text{·}585$ volt.

If the direction of the arrow be from electrolyte to
electrode (always within the element), the electrolyte is
negative to the electrode ; otherwise it is positive.

As is evident, the electromotive force of single cells
consists in general of two potential differences, which
may be in the same direction, their sum constituting
the electromotive force of the cell, or they may be
oppositely directed, when they partially neutralise each
other. The first of the three examples corresponds
to the former case, and the two others to the latter.
Of the accompanying figures the first graphically
represents the changes of potential for the three closed
elements under the assumption that the external

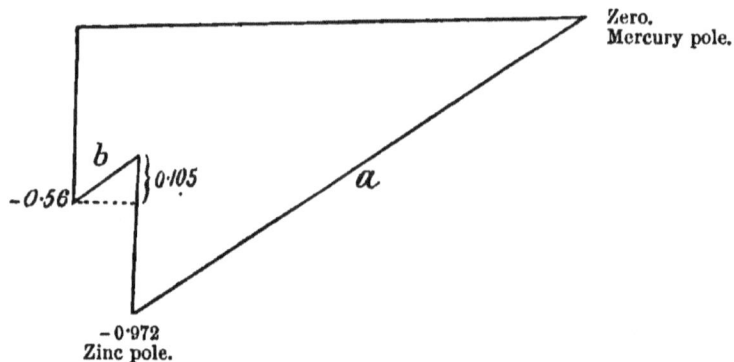

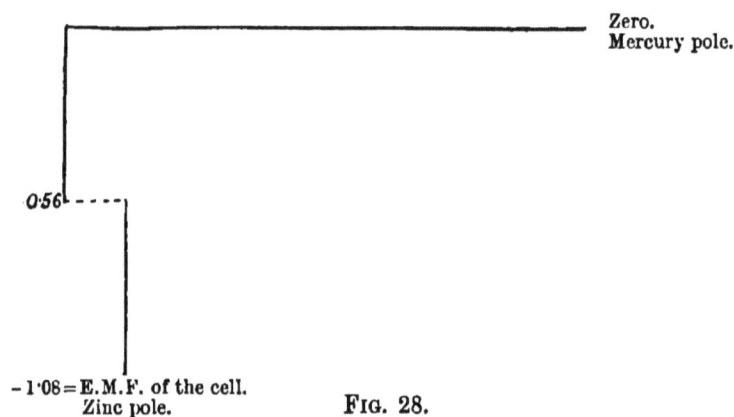

Fig. 28.

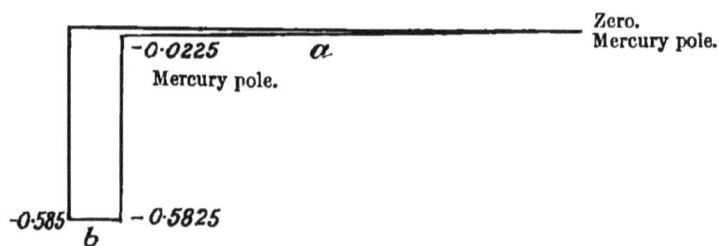

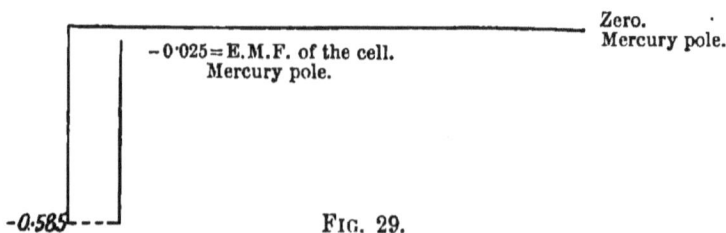

Fig. 29.

resistance (a) and the internal (b) are the same per unit length, and that (a) as a whole is nine times as great as (b).   The three other figures illustrate the potential differences when the cells are open.   In the first two the positive, and in the last two the negative poles are connected with the earth.

The following explanations apply to Fig. 30 (see also p. 13).   If the mercury of the cell be brought to the potential zero, the potential of the potassium chloride solution is evidently -0.56 when the circuit is open.   Therefore, when the circuit is closed the potential of that layer of the electrolyte in immediate contact with the mercury is given, and is indicated in the figure by the perpendicular at A.   At the point B, or the place of contact between electrolyte and oxygen, there is a potential difference of 1.31 volt, and the electromotive force of the whole cell is 0.75 volt. That is, with open circuit, when the mercury electrode is connected with the earth, the oxygen electrode has a potential of 0.75 volt.   When the circuit is closed and the internal resistance of the cell is one-tenth of the total, the oxygen electrode indicates a potential of but $0.75 \times \frac{9}{10}$, or 0.675 volt ; consequently a perpendicular must be erected at B corresponding to this 0.675 volt, and a layer of the electrolyte in the immediate neighbourhood must have a potential of -0.635, since the electrolyte is negative towards the electrode and the potential difference is 1.31 volt. The perpendicular at B must therefore be continued below the line AB to a point corresponding to -0.635 volt.   It is now only necessary to draw the lines (a) and (b), representing the fall of potential corresponding to the resistance of the circuit, to complete the figure for the closed circuit.

The cases illustrated by the other figures are analogous, and their consideration is recommended to the student.

Neumann [1] determined the following potential

+0·675 Oxygen pole.

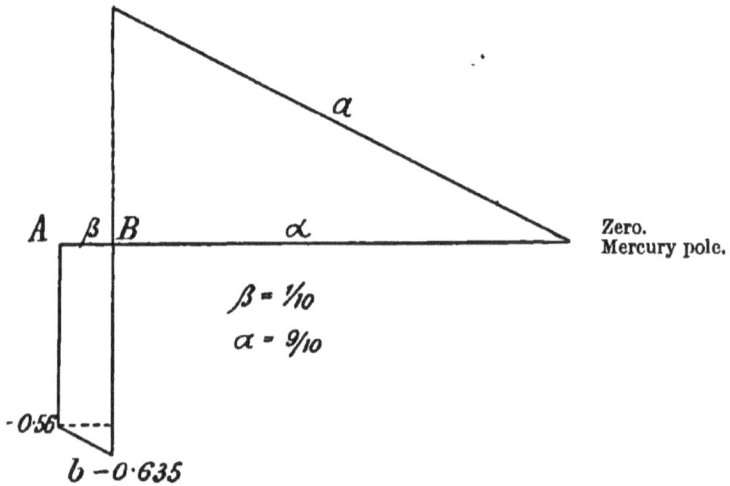

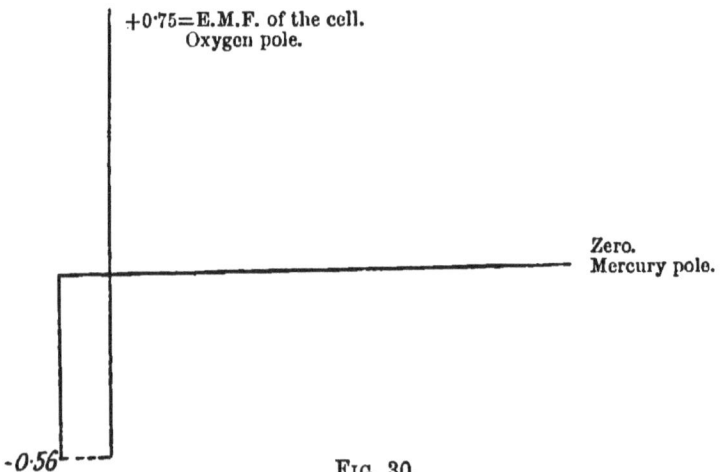

α

A β B α Zero.
Mercury pole.

$\beta = \frac{1}{10}$

$\alpha = \frac{9}{10}$

-0.55

$b -0·635$

+0·75 = E.M.F. of the cell.
Oxygen pole.

Zero.
Mercury pole.

-0.56

FIG. 30.

differences for the metals in normal or saturated solutions of their salts.

[1] *Zeitschr. physik. Chem.* xiv. 229, 1894.

Metal.	Sulphate.	Chloride.	Nitrate.	Acetate.
Magnesium .	+ 1·239	+ 1·231	+ 1·060	+ 1·240
Aluminium .	+ 1·040	+ 1·015	+ 0·775	...
Manganese .	+ 0·815	+ 0·824	+ 0·560	...
Zinc . . .	+ 0·524	+ 0·503	+ 0·473	+ 0·522
Cadmium . .	+ 0·162	+ 0·174	+ 0·122	...
Thallium . .	+ 0·114	+ 0·151	+ 0·112	...
Iron . . .	+ 0·093	+ 0·087	...	...
Cobalt . .	− 0·019	− 0·015	− 0·078	− 0·004
Nickel . .	− 0·022	− 0·020	− 0·060	...
Lead . . .	...	− 0·095	− 0·115	− 0·079
Hydrogen . .	− 0·238	− 0·249	...	− 0·150
Bismuth . .	− 0·490	− 0·315	− 0·500	...
Arsenic . .	...	− 0·550	...	...
Antimony .	...	− 0·376	...	...
Tin . . .	...	− 0·085	...	...
Copper . .	− 0·515	...	− 0·615	− 0·580
Mercury .	− 0·980	...	− 1·028	...
• Silver . .	− 0·974	...	− 1·055	− 0·991
Palladium . .	...	− 1·066	...	...
Platinum . .	...	− 1·140	...	...
Gold . . .	...	− 1·356	...	...

The values for bismuth, arsenic, antimony, and tin are not comparable with the others, as the corresponding solutions contained free acid, and nothing is certainly known regarding the quantity of ions contained in such solutions. They were made by dissolving one equivalent in grams of the solid substance in a liter of water, the resulting precipitate being removed by filtration. Because of undetermined conditions in the cases of the gold chloride and the hydro-chloroplatinic acid solutions, the values for these metals cannot be considered fixed. Nothing is known in these cases concerning the numbers of ions present. Finally, the values of magnesium, aluminium, and manganese, the water-decomposing metals, are only to be considered as lower limits, their action upon the water causing the values of cells containing them to diminish immediately after introduction of the electrode.

For the remaining solutions the numbers of ions in the corresponding electrolytes were approximately the same ; they are, however, not identical, as the solutions were by no means completely dissociated.   In order to make them perfectly comparable, *i.e.* to give them all the same ion concentration (it being upon this that the magnitude of the potential difference depends), it would be necessary to take into consideration the degree of dissociation in each case.   The values as given, however, suffice for comparison.   The order of the electromotive forces, as the formula

$$\pi = \frac{RT}{n_e \epsilon_0} \ln \frac{P}{p}$$

shows, where p has nearly the same value for all the electrolytes, presents also the order of the solution pressures (P) of the various elements.   This is then the actual electromotive series of the metals.

Influence of Negative Ions upon the Potential Difference :   Metal — Metal-Salt Solution. — The question may still be asked : Is the nature of the negative ion without influence upon the potential difference ?   This cannot be surely answered from the above values for chloride, sulphate, and acetate. Differences occur in these cases from differences in degree of dissociation in the individual solutions.   But since the degrees of dissociation are not known with sufficient certainty, it cannot be determined whether the differences of the dissociation degrees completely explain the irregularities or not.   Neumann (*l.c.*) consequently prepared 0·01 normal solutions of over twenty different thallium salts (mostly salts of organic acids), and determined the potential difference when these are in contact with metallic thallium.   In these solutions the salts may be considered as equally dissociated, and the same potential differences might be expected in each case. As the measured values do not differ by more than 0.001 volt, the conclusion is justified that the nature of the negative ion is without influence upon the potential of the metal. Negative ions by which the metal is chemically affected—as, for instance, NO_3—are, of course, excluded from this generalisation. This explains why the nitrate solutions cause very different potentials from the chloride, notwithstanding a nearly equal dissociation.

Electrolytic Solution Pressure.—The magnitudes of the electrolytic solution pressures of the metals may be ·directly ascertained from the above measurements. The potential difference at the electrode is

$$\pi = \frac{RT}{n_e \epsilon_0} \ln \frac{P}{p},$$

and since the values of π and p are known, P is calculable. If p be expressed in atmospheres, P is obtained in the same unit.

Assuming the osmotic pressure in the totally dissociated normal solution to be 22 atmospheres, Neumann [1] obtained the following values for P. Special attention was given to the degree of dissociation at the ordinary temperature $(17°)$.

Zinc	.	.	$= 9.9 \times 10^{18}$	Atmospheres.
Cadmium	.	.	$= 2.7 \times 10^6$	,,
Thallium	.	.	$= 7.7 \times 10^2$	,,
Iron	.	.	$= 1.2 \times 10^4$	,,

---

[1] Newmann calculated the values from the incorrect formula

$$\pi = \frac{RT}{n_e \epsilon_0} \left(\ln \frac{P}{p} - 1 \right),$$

and they have consequently been corrected.

Cobalt .	.	.	$= 1·9 \times 10^0$ Atmospheres.
Nickel .	.	.	$= 1·3 \times 10^0$,,
Lead .	.	.	$= 1·1 \times 10^{-3}$,,
Hydrogen	.	.	$= 9·9 \times 10^{-4}$,,
Copper	.	.	$= 4·8 \times 10^{-20}$,,
Mercury	.	.	$= 1·1 \times 10^{-16}$,,
Silver .	.	.	$= 2·3 \times 10^{-17}$,,
Palladium	.	.	$= 1·5 \times 10^{-36}$,,

This may be considered as *the absolute electromotive series of the metals*. Each metal, when placed in a solution of one of those following, causes the precipitation of the latter or the evolution of hydrogen. It has already been seen that hydrogen ions are present in pure water, and accordingly also in the solution of any substance. Whether hydrogen is generated or not depends upon which of the two positive ions, the hydrogen or the metal, changes more easily into the non-electric condition. Hydrogen ions can be continually generated from the undissociated water.

A word of explanation may be added concerning the hydrogen. In considering the concentration cells (p. 163) it was seen that the solution pressure of the hydrogen depends upon its gas pressure (and this also applies to other gases). The electrode material, platinised platinum, does not come into account. The above value for hydrogen is that for atmospheric pressure. Theoretically, the solution pressure of the hydrogen may be increased or diminished as desired by altering the pressure under which it is confined. The limits in both directions are, however, practically soon reached. According to page 166, the gas pressures vary as the squares of the solution pressures. If, for example, hydrogen under a pressure of 10,000 atmospheres be used, its solution pressure is relatively little changed: it becomes $9·9 \times 10^{-2}$ instead

of $9{\cdot}9 \times 10^{-4}$ atmospheres.   Such   pressures   can
scarcely be attained.

In agreement with the table, it has been shown
that platinum black, charged with hydrogen at atmospheric pressure, is capable of precipitating the metals
following it from their salt solutions.   If ordinary
platinum were used, the process would require a very
long time, because of its slight solvent action.

Influence of Dilution.—The essential points concerning the reversible cells, such as the Daniell, where
the ion-producing substances are elements, have already
been treated.   The effect of dilution upon the electromotive force of an element will next be considered
because of its importance.   The electromotive force,
when the electrodes are capable of producing only
negative or positive ions, is given by the equation
(p. 208).

$$\pi = \frac{RT}{n_{\bullet}\epsilon_0} \ln \frac{P}{p} - \frac{RT}{n'_{\bullet}\epsilon_0} \ln \frac{P'}{p'}.$$

As is evident from the formula, an increase in the concentration of the one solution diminishes the electromotive force, and of the other increases it.   In the
Daniell element, for example, the electromotive force
is increased by concentrating the copper sulphate solution, and decreased by concentrating the zinc sulphate.

For both kinds of cells it may be said that
concentration of the solution from which the ions
separate causes an increase, while concentration of that
in which new ions are produced causes diminution of
the potential difference.   This is easy to comprehend
when it is remembered that the osmotic pressure
opposes the solution pressure.   In the first case the
passage of the ions from the solution is made easier,

and the electromotive force increases; in the second their entrance is made more difficult, and the electromotive force diminishes. If the two metals of the electrodes have the same valency, equivalent changes in the concentrations of the two solutions do not affect the electromotive force.

If, finally, one electrode produces negative and the other positive ions, the following holds:

$$\pi = \frac{RT}{n_e \epsilon_0} \ln \frac{P}{p} + \frac{RT}{n'_e \epsilon_0} \ln \frac{P'}{p'}.$$

In this class of cells an increase in the concentration of either solution causes a diminution of the electromotive force, since ions are simultaneously produced at both electrodes, and the increased osmotic pressure opposing the introduction of ions reduces the electromotive force. The magnitude of the changes of electromotive force, produced by given alterations in the concentrations, may be recognised from p. 170.

A single exception to this generalisation is the gas cell: Platinised platinum in hydrogen, electrolyte, e.g. sulphuric acid solution—platinised platinum in oxygen. The above formula applies also to this cell, p and p' being the osmotic pressures of the hydrogen and hydroxyl ions. But as the product of the concentration of these ions in pure water, or in any aqueous solution, always has the same value, the electromotive force of the cell cannot be altered by changing the concentration of the electrolytes, nor in general by changing the electrolytes themselves. If p increases to a certain extent above its original value, p' diminishes to the corresponding degree. This is true so long as the two electrodes are in contact with solution homogeneous as regards the H and OH ions, for the electromotive force only depends upon the layers of electrolyte at the electrodes.   If the electrodes are originally placed in different solutions, or if the portions about the electrodes become altered during the passage of the current, as when a salt solution is used for electrolyte, acid appearing at one electrode and base at the other, this cell may be included in the ordinary class.

Heat of Ionisation.—The Helmholtz formula,

$$\epsilon_0 \pi - Q = \epsilon_0 T \frac{d\pi}{dT},$$

is applicable not only to the whole, but also to each individual potential difference in the cell.   Q then represents the heat effect produced at the electrode in question, and $\frac{d\pi}{dT}$ the temperature coefficient of the potential difference.   The electromotive force of the cell consisting of two or more single potential differences, the temperature coefficient is composed of their individual temperature coefficients.

If, for example, the potential difference between zinc and zinc sulphate solution be known, and its temperature coefficient determined, the value of Q may be calculated. This is the heat generated by the passage of metallic zinc into the ionic condition, that is, the heat of ionisation of the zinc.   The thermo-chemical data are always sums or differences of two or more of these heats of ionisation.   The precipitation of copper from its solution by zinc gives the difference between the heats of ionisation of zinc and copper.   On the other hand, if the heat of ionisation for a single element be known, as here the zinc, that of the others may be obtained from the thermo-chemical data.   The following table containing the heats of ionisation is given by

Ostwald.[1] Because of uncertainty of some of the experimental data, the values are only approximately correct. K is very nearly equal to a hundred small calories.

	For one Atomic Weight.	For one Equivalent Weight.
Potassium	+ 610 K	610 K
Sodium	+ 563 „	563 „
Lithium	+ 620 „	620 „
Strontium	+1155 „	578 „
Calcium	+1070 „	535 „
Magnesium	+1067 „	534 „
Aluminium	+1175 „	392 „
Manganese	+ 481 „	240 „
Iron (ferrous ions)	+ 200 „	100 „
„ (ferrous ions in ferric ions)	− 121 „	− 121 „
Cobalt	+ 146 „	+ 73 „
Nickel	+ 135 „	68 „
Zinc	+ 326 „	163 „
Cadmium	+ 162 „	81 „
Copper (cupric ions)	− 175 „	88 „
„ (cuprous ions)	− 170 „ (?)	− 170 „ (?)
Mercury	− 205 „	− 205 „
Silver	− 262 „	− 262 „
Thallium	+ 10 „	+ 10 „
Lead	− 10 „	− 5 „
Tin	+ 20 „	+ 10 „

Direct Measurement of both Potential Differences in a Cell.—Instead of measuring one of the potential differences and determining the other by subtraction, it is possible to measure the potentials singly. The difference at the point of contact of the two liquids being reduced as much as possible, the sum of the two measured potential differences must be very nearly equal to the electromotive force. Rothmund[2] corroborated this, which again proves that no considerable

---

[1] *Zeitschr. physik. Chem.* xi. 501, 1893.
[2] *Ibid.* xv. 1, 1884.

potential differences exist at the contact points between metals.

For determining the single potential differences, Rothmund made use of the Lippmann method already described. For mercury he substituted amalgams of the baser metals, which, in moderate concentration (about ·01 per cent), act very nearly as the pure metals. He determined, for example, the potential difference when lead amalgam is in contact with normal sulphuric acid solution saturated with lead sulphate, copper amalgam in contact with normal sulphuric acid solution containing ·01 molecular weight in grams of copper sulphate per liter, and constructing cells with the normal electrode whose value is directly determined, he measured the resulting electromotive force, and compared it with the sum of the known values for the two single potential differences. In order to reduce the magnitude of the potential difference at the contact surfaces between the solutions, the combination, mercury — mercurous sulphate in normal sulphuric acid was used instead of the normal electrode described. This combination will be represented by N′, and its potential difference is − 0·926, being greater than that of the other normal electrode, because mercurous sulphate is more soluble than the chloride.

The following values were obtained :

Copper amalgam—normal sulphuric acid,
    with ·01 mol. $CuSO_4$ per liter.   .   = − 0·445 V.
N′   .   ′   .   .   .   .   = − 0·926 „
Lead amalgam—normal sulphuric acid
    saturated with lead sulphate .   .   = + 0·008 „

The electromotive force of the copper-N′ cell should therefore be 0·481 volt, and that of the lead amalgam-

N′ cell 0·918 volt. The experimentally determined values are 0·458 and 0·926. It is therefore impossible that greater potential differences can exist between the metals themselves than the differences between these values.

CELLS IN WHICH THE ION-PRODUCING SUBSTANCES ARE NOT ELEMENTS

A class of chemical cells, apparently very different from that represented by the Daniell element, will now be considered. If a platinised platinum electrode is surrounded by a solution of stannous chloride, and another by one of ferric chloride, and the two are placed in metallic connection, an electric current is obtained, which passes within the cell from the former solution to the latter. The trivalent ferric ions give up an equivalent of electricity, becoming ferrous ions, while each stannous ion takes up two electrical equivalents, becoming a stannic ion. The process may be imagined in detail as follows: The stannous ions change into stannic, and thereby positive electricity is produced. Since this can never come into existence alone in a change of chemical into electrical energy, electricity must be produced upon the electrode. This electricity passes through the wire to the other electrode, where it unites with the positive electricity derived from the change of ferric into ferrous ions.

The cell, platinised platinum in hydrogen, electrolyte A, electrolyte B, platinised platinum in chlorine, is evidently completely analogous to the above combination. It was previously stated (p. 163) that platinised platinum in hydrogen may be considered as a hydrogen electrode. In a similar manner the above combination may be characterised as stannous and ferric electrodes, and just as a tendency to go into the ionic (or of the ions to go into the neutral) state was ascribed to the hydrogen and chlorine electrodes, so a tendency of the stannous and ferric to form stannic and ferrous ions may be recognised. The electromotive force of this cell also consists principally of the two independent potential differences occurring at the electrodes. But these potential differences depend not only upon the solution pressures of the substances in question, but also upon the osmotic pressures of the ions forming. Therefore the concentrations of the stannic ions formed at the one electrode, and of the ferrous ions at the other, are important factors; a certain constant potential difference, as in the Daniell element, could only be expected when the solutions already contained stannic and ferrous ions. Moreover, the concentration of the altering compounds must be considered, for the solution pressure of a substance at constant temperature is invariable only at a definite concentration.

From what has been said, it is obvious that there is essentially no difference between the Daniell and the so-called reduction and oxidation cells. The laws governing the former may be expected to control the latter.

Experimental investigation has not been carried out sufficiently to demonstrate the accuracy of all the theoretical deductions. Thus the influence of the concentration of the substances formed at the electrodes has been almost entirely neglected, and it is probable that the varying values of such cells are due to this. The non-reversibility of these cells may be similarly accounted

for.   If, instead of allowing the stannous chloride—
ferric chloride cell to act, it be opposed by a cell of
greater electromotive force, oxygen must separate at
one electrode (at least in dilute solution) and metallic
tin at the other.   Stannic and ferrous chlorides being
present, a change of the stannic into the stannous,
and of ferrous into ferric salt, would certainly
take place instead of the above, and the cell be
reversible.[1]

A cell whose electrodes are zinc and chlorine, and
whose electrolytes do not contain zinc and chlorine
ions, is no longer a reversible cell.   If a stronger
opposing current be sent through such a cell, the positive ions of one electrolyte separate at the zinc, and
the negative of the other at the chlorine electrode,
while zinc and chlorine ions are liberated through its
own activity as a cell.

Bancroft proved that the electromotive force of
such cells is essentially the sum of the two single
potential differences.

Although our knowledge of the values of such
quantities leaves much to be desired, the following
list of potential differences, including the elements,
chlorine, bromine, and iodine, is given, it being not
only of considerable interest, but presenting, in addition, a measure of the "strength" of the substances.
The following values were obtained from platinised
electrodes surrounded by the liquids mentioned.   Most
of the solutions contained about $\frac{1}{5}$ molecular weight in
grams per liter :[2]

---

[1] It is very doubtful if the processes even of most of such cells
are practically capable of proceeding reversibly.   At any rate, this
circumstance complicates the relations.

[2] *Zeitschr. physik. Chem.* xiv. 193, 1894.

$SnCl_2 + KOH$ . . $+ 0.301$	Hydroxylamine . $- 0.636$
Na_2S . . . $+ 0.091$	$NaHSO_3$. . . $- 0.663$
Hydroxylamine, $KOH + 0.056$	H_2SO_3 . . . $- 0.718$
Chromous acetate,	$FeSO_4 + H_2SO_4$. $- 0.794$
KOH . . . $+ 0.029$	Potassium ferric
Pyrogallol, KOH . $- 0.078$	oxalate . . $- 0.846$
Hydrochinone, KOH $- 0.231$	$I_2 - KI$ . . . $- 0.888$
Hydrogen, HCl . $- 0.249$	K_3FeCy_6. . . $- 0.982$
Potassium ferrous	$K_2Cr_2O_7$ . . . $- 1.062$
oxalate . . $- 0.285$	KNO_3 . . . $- 1.137$
Chromous acetate . $- 0.364$	$Cl_2 - KOH$ . . $- 1.186$
K_4FeCy_6, KOH . $- 0.474$	$FeCl_3$ . . . $- 1.238$
I_2, KOH. . . $- 0.490$	HNO_3 . . . $- 1.257$
$SnCl_2 - HCl$ . . $- 0.496$	$HClO_4$ . . . $- 1.267$
Potassium arseniate . $- 0.506$	$Br_2 - KOH$ . . $- 1.315$
NaH_2PO_2 . . $- 0.516$	$H_2Cr_2O_7$ . . . $- 1.397$
$CuCl_2$ . . . $- 0.560$	$HClO_3$ . . . $- 1.416$
$Na_2S_2O_3$ . . . $- 0.576$	$Br_2 - KBr$ . . $- 1.425$
Na_2SO_3 . . . $- 0.583$	KIO_3 . . . $- 1.489$
Na_2HPO_3 . . $- 0.593$	$MnO_2 - KCl$ . . $- 1.628$
K_4FeCy_6. . . $- 0.595$	$Cl_2 - KCl$ . . $- 1.666$
$FeSO_4$ (neutral) . $- 0.633$	$KMnO_4$ . . . $- 1.763$

In electrical processes the so-called oxidations and reductions may be clearly distinguished, for it may be said that the process is always one of oxidation when negative electricity is produced on an ion or positive disappears. When positive electricity appears or negative disappears the process is one of reduction. According to these definitions there must be, in every galvanic element, an oxidation at one electrode and a reduction at the other. In the Daniell element the reduction takes place at the zinc electrode, and the oxidation at the copper. The precipitation of one metal by another, the process of substitution, is thus to be considered as one of oxidation and reduction. It is evident, then, that the metals can only serve as reducing agents, since they are only capable of producing positive ions.

The negative elements, on the other hand, or the substances producing negative ions, act exclusively as oxidising agents. Salt solutions in general may be reducing as well as oxidising agents, for they contain both positive and negative ions, and are therefore capable of yielding positive and negative electricity. If zinc be placed in a solution of cadmium bromide, cadmium is precipitated, the solution acting as an oxidising agent; but if chlorine be conducted into the solution, bromine separates, the solution acting as reducing agent.

Similarly, the substances in the above table may be examined to discover whether they are reducing or oxidising agents. From the above it is, moreover, not surprising that a dissolved substance may have a reducing or oxidising action according to circumstances. This may even be the case when only the single ion enters the reaction; the bivalent ferrous ion may change into the trivalent ion, on the one hand, or into metallic iron, on the other, that is, it may act reducing or oxidising.

In reversible cells each electrode may be made the seat of the oxidation or reduction at will.

It may be well to say a word here concerning the conditions which determine the actual production of the electric current.[1] It has been seen that in all galvanic elements a reduction and oxidation take place, that is, at one electrode ions come into existence, and disappear at the other. *That the reaction may be the source of an electric current, the two processes must take place at points separated from each other.* If they both occur at the same point, no electric current

---

[1] Ostwald, *Chemische Fernewirkung. Zeitschr. physik. Chem.* ix. 540, 1892.

results. Zinc being placed in a copper sulphate solution, both the oxidation and reduction proceed simultaneously at the surface of the metal. The electric charges of the dissolving zinc and precipitating copper have the opportunity of neutralising each other there, and the possibility of a removal of this neutralisation to some other point (and thereby the production of an electric current) is lost. Hence the statement that a chemical reaction between two substances can only be used as a source of the electric current when electricity is produced or disappears in the reaction (*i.e.* by changes in the charges of the ions), and also when the two substances separated from each other are still capable of undergoing this reaction.

If zinc be in contact with a solution of zinc sulphate, and a platinum wire be placed therein, no current is obtained on connecting the wire with the zinc. If it be desired to dissolve the zinc, that is, to cause it to pass into the ionic state and produce a current, this may be accomplished by surrounding the *platinum* with a solution of a copper salt, or an acid whose positive component has a smaller tendency to produce ions than zinc. The addition of the copper or acid solution directly to the zinc solution would evidently not produce an electric current.

In the production of galvanic currents many different oxidising agents have been used to achieve the highest possible efficiency, without the theory of the phenomena being clearly understood. One of the most common cells is the bichromate element consisting of zinc — chromic acid (or sodium bichromate and sulphuric acid) — carbon. The process essentially consists in the formation of zinc ions at the negative (zinc) electrode, and the reduction of chromium ions at the positive (carbon) electrode from higher to lower valency, whereby electricity is given up to the electrode.

The electromotive force of this cell is great, because the zinc has a strong tendency to go into the ionic state, and the chromium ions of high valency also tend strongly to change into ions of lower valency, the two tendencies additively producing the high electromotive force.   Furthermore, it is clear that the electromotive force of this cell, when active, must gradually diminish, because zinc ions are continually forming, while the concentration of the chromium ions of higher valency is decreasing, and that of those of lower valency increasing.   Each of the three changes reduces the electromotive force.

The energetic oxidation of the zinc and the high electromotive force of the cell is therefore obtained by the addition of the oxidising agent not to the zinc but to the carbon.

It is also possible to dissolve the noble metals or change them into ionic state in a similar manner.   A cell consisting of platinum — sodium chloride solution — gold, produces no electric current, though one is produced when chlorine water is introduced at the platinum electrode, the gold dissolving.   The great tendency of the chlorine to yield ions may be looked upon as forcing the resisting gold to act similarly. Addition of the chlorine water to the gold electrode alone would not result in the production of a current (the platinum being unaffected), and the gold would oxidise very slowly.

POTENTIAL DIFFERENCE BETWEEN SOLID AND LIQUID METALS AT THE MELTING POINT

Before ending the chapter on reversible cells a special case may receive brief mention. Imagine two electrodes of the same metal in contact with an electrolyte at the melting point of the metal, and suppose one of the electrodes to be liquid and the other solid; would such a cell produce an electromotive force? For instance, would the current pass from the liquid to the solid electrode, and perhaps the heat of fusion be the source of the resulting electrical energy? The impossibility of such a process may be easily grasped. Suppose electrical energy could be produced by such an arrangement, and that all the material has passed from the liquid to the solid electrode through the action of the cell. This being then melted by the application of heat from the surroundings at the constant temperature of its melting point, a current could be produced in the direction opposite to the first, and so on. In this way heat of constant temperature would perform any desired amount of work in a cyclic process, which is contrary to the second law of thermodynamics.

CHAPTER VII

POLARISATION

THE phenomena observed when an electric current is conducted through an electrolyte between inactive electrodes, as gold, platinum, carbon, etc., will now be considered. It has long been known that the current produces a decomposition of the electrolyte at the electrodes, and that its electromotive force is thereby reduced. The two facts are evidently related. The performance of an amount of work, more or less considerable according to circumstances, is necessary to bring about the decomposition of an electrolyte (as, for example, hydrochloric acid into hydrogen and chlorine), and this work is done by the electric current. When such reduction of the electromotive force occurs, polarisation is said to take place. The phenomenon was formerly very little understood, and it is only within the last few years that its explanation has become possible.

If a current flows for a time through the above-described arrangement, and is then interrupted, the two electrodes being connected through a galvanometer, it will be observed that an electric current, which rapidly becomes weaker, passes between the electrodes in a direction opposite to that of the first or applied current. This is spoken of as the " polarisation current," and its electromotive force is called the " electromotive force of polarisation." From the following it will be evident that this current is derived from the tendency of the material separated in the neutral condition to return to the ionic condition.

Ohm's law, applied to a circuit possessing a certain primary electromotive force π_1, and containing a " polarisation cell," is represented by

$$C = \frac{\pi_1 - \pi_2}{R},$$

where π_2 is the electromotive force of polarisation, C the current-strength, and R the total resistance of the circuit.

Method of measuring Polarisation.—As already seen, the electromotive force of polarisation is not a constant, but rapidly diminishes when the primary electromotive force is removed; its magnitude is therefore best determined during the passage of the primary current. The accompanying figure represents an arrangement which may be used for the measurement.

One circuit is represented by 1, 2, a, 1, and the other by 2, e, b, a, 2; 1 is the source of the electricity, 2 the polarisation cell, e a compensation electrometer, b a known electromotive force, which may be altered at will, and a a tuning-fork commutator, which vibrates very rapidly. The arrangement is such that at a one circuit is opened and the other simultaneously closed, then the latter opened and the former closed, etc., with each vibration of the tuning-fork. The result is practically the same as though both primary and polarisation current were independently active. Thus the electromotive force of the

latter may be measured under the same conditions as
if the primary circuit were continually closed.   It is
only necessary to alter b until the electrometer

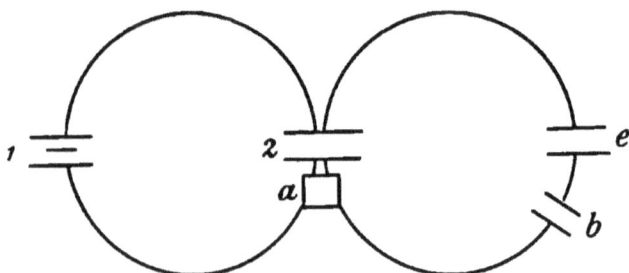

Fig. 31.

shows a condition of equilibrium; b is then the
desired value.

As the electromotive force of galvanic elements is
due to two or more potential differences, so also in the

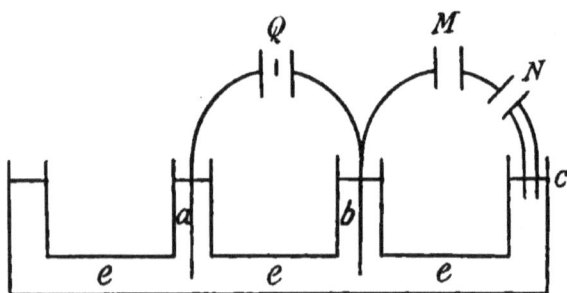

Fig. 32.

electromotive force of polarisation two single potential
differences are found [1] located at the two electrodes.
In order to measure them separately, the method of
Fuchs is employed.   Its arrangement is shown in
Fig. 32.   A double U tube is filled with the solution
of the electrolyte (e), whose polarisation is to be

[1] The fall of potential due to the resistance of the electrolytes is
avoided by the method.

measured. a and b are two indifferent electrodes connected with the source (Q) of the primary or polarising current. If the potential difference at b is to be measured, the bent glass tube of the normal electrode (N) (p. 217), filled with normal potassium chloride solution, is inserted at c in the electrolyte (e), and b is connected with the mercury of the normal electrode by means of the platinum wire of the latter. An element thereby results, consisting of two electrodes and two electrolytes, and the electromotive force of the combination is measured by the usual apparatus at M. The potential difference between b and e may then be determined by subtraction of the normal electrode potential, and that at the surface of contact between the liquids from the total electromotive force. For determining the potential difference between a and e the process is analogous, and using a primary or polarising current, whose electromotive force gradually increases from zero, it is observed that the electromotive force of polarisation is at first very nearly equivalent to that of the primary current. As the latter becomes higher the former falls gradually away from it in magnitude, nevertheless always increasing to some extent. The much-sought-after maximum of polarisation does not actually exist.

Decomposition Values of the Electromotive Force. —There is another characteristic point for the different electrolytes. A continuous current and continuous decomposition only take place when the electromotive force exceeds a certain value. If an electromotive force less than the above be inserted, only an instantaneous passage of electricity takes place, which may be made evident by a galvanometer in the circuit.

The needle of the galvanometer is at first deflected, but returns very nearly to its original position (the effect of secondary disturbing influences will be considered later), which does not happen when the applied electromotive force has reached the value in question.

Le Blanc determined the magnitudes of these decomposition values for a great many electrolytes, chiefly in normal solutions. They may be very exactly determined for salts from which a metal is precipitated by the current, but for other salts, as well as for acids and alkalies, they are less easily found. The following decomposition values were found for salts from which the metal is precipitated.[1]

$ZnSO_4$	$= 2\cdot35$ Volts		$Cd(NO_3)_2$	$= 1\cdot98$ Volts
$ZnBr_2$	$= 1\cdot80$	,,	$CdSO_4$	$= 2\cdot03$	,,
$NiSO_4$	$= 2\cdot09$	,,	$CdCl_2$	$= 1\cdot88$	,,
$NiCl_2$	$= 1\cdot85$	,,		
$Pb(NO_3)_2$	$= 1\cdot52$	,,	$CoSO_4$	$= 1\cdot92$	,,
$AgNO_3$	$= 0\cdot70$	,,	$CoCl_2$	$= 1\cdot78$	,,

The decomposition values for sulphates and nitrates of the same metal, as shown by the experiments with cadmium salts and other experiments with the alkalies, are nearly equal. As is evident, the values for the various metals are different. The conclusion to be drawn from the corresponding values for the acids and bases is that there exists a maximum decomposition point, which is exhibited by most of the compounds and exceeded by none. This is about $1\cdot70$ volt. Among the acids, however, several gave various values below this maximum. The following tables contain the values for acids and bases:

[1] *Zeitschr. physik. Chem.* viii. 299, 1891.

Acids

Sulphuric	= 1·67	Volt
Nitric	= 1·69	,,
Phosphoric	= 1·70	,,
Monochloracetic	= 1·72	,,
Dichloracetic	= 1·66	,,
Malonic	= 1·69	,,
Perchloric	= 1·65	,,
Dextrotartaric	= 1·62	,,
Pyrotartaric	= 1·57	,,
Trichloracetic	= 1·51	,,
Hydrochloric	= 1·31	,,
Hydrazoic	= 1·29	,,
Oxalic	= 0·95	,,
• Hydrobromic	= 0·94	,,
Hydroiodic	= 0·52	,,

Bases

Sodium hydrate	= 1·69	Volt
Potassium hydrate . . .	= 1·67	,,
Ammonium hydrate . . .	= 1·74	,,
$\frac{1}{4}$ n. Methylamine	= 1·75	,,
$\frac{1}{2}$ n. Diethylamine . . .	= 1·68	,,
$\frac{1}{8}$ n. Tetramethyl ammonium hydrate	= 1·74	,,

The alkali and alkaline earth salts of the strongly dissociating acids with maximum decomposition values, as sulphates and nitrates, have nearly the same decomposition point—about 2·20 volts. The chlorides, bromides, and iodides have lower values, independent of the nature of the alkali metal. Additivity is exhibited owing to the mutual independence of the potential differences produced at the two electrodes. Differences between the values for the various halogen compounds of the alkalies, hydrogen, and the metals are nearly equal ; for example, the difference between hydrochloric and hydrobromic acid is the same as that between sodium chloride and bromide.

The salt of a slightly dissociating acid, as sodium acetate, or of a slightly dissociating base, as ammonium sulphate, always exhibits a lower value than that of a highly dissociating acid or base, presupposing that the acid and base possess the maximum decomposition value. The halogen salts of ammonium have lower decomposition values than the corresponding salts of the alkalies ; and, in fact, the differences between corresponding salts are equal.

Concerning the effect of dilution in the case of bases and acids which on electrical decomposition evolved oxygen and hydrogen at the electrode, the decomposition values were independent of the dilution, and this is true for all the acids excepting those whose decomposition values are below the maximum. For these the value rises with increasing dilution, and finally reaches the maximum. This is very marked in the case of hydrochloric acid.

Decomposition Point.

$\frac{2}{1}$ Normal hydrochloric acid,		1·26 volt
$\frac{1}{2}$,,	,,	1·34 ,,
$\frac{1}{6}$,,	,,	1·41 ,,
$\frac{1}{16}$,,	,,	1·62 ,,
$\frac{1}{32}$,,	,,	1·69 ,,

It is also worthy of note that when the maximum value is reached, the acid solution is no longer decomposed into chlorine and hydrogen, but into hydrogen and oxygen.

The above experiments were carried out with platinum electrodes. If other electrodes be used, as gold or carbon, different numerical values are obtained, but the general relations between them remain unaltered.

In order to obtain a better insight into polarisation phenomena, Le Blanc[1] investigated the potential difference at the electrode, where the metal is electrolytically deposited (the cathode), when the electromotive force of the primary current is gradually increased from zero. The result of this investigation is that the potential difference at the decomposition point was found to be equal to that which the precipitating metal would itself exhibit in the solution. For example, a normal solution of cadmium sulphate was decomposed at a primary electromotive force of $2\cdot03$ volts. The potential difference of the electrode where the cadmium separated was $+0\cdot16$ volt with regard to the electrolyte. Metallic cadmium placed in the solution also gave $0\cdot16$ volt. In many solutions the electrode exhibited the potential difference due to the separating metal before the decomposition point of the solution was reached. For instance, in silver nitrate the electrode had the value of pure silver in silver nitrate even below the decomposition point $(0\cdot70)$. This is due to the great tendency of the silver ions to separate as electrically neutral metal.

It could also be demonstrated that the material of the indifferent electrodes is without influence upon the magnitude of these potential differences. The results were the same whether gold, platinum, carbon, or any other metal more negative than that in solution was used. From this it is evident that the electrode itself · possesses no "specific attraction" for the electricity, as formerly imagined.

The process of precipitation and solution of the metals is, therefore, to be considered irreversible. It may be represented as follows. If an indifferent

[1] *Zeitschr. physik. Chem.* xii. 333, 1893.

electrode be placed in the solution of a metallic salt, a very small quantity of the ions must leave the ionic state and be deposited upon the electrode as metal ; for if the electrode contained absolutely none of the metal of the salt solution, the potential difference between electrode and electrolyte would be infinitely great, in accordance with the formula, which is also applicable to polarisation phenomena,

$$\pi = \frac{RT}{n_e \epsilon_0} \ln \frac{P}{p}.$$

If $P = 0$, π must be infinite, but π must always have a finite value (otherwise an infinite amount of work could be done) ; therefore it must be assumed that upon the electrode, and also in the solution, there are always traces of the metal. The metal must, therefore, separate upon the electrode until the tendency of the ions to precipitate is exactly compensated by the electrostatic attraction due to the electrode becoming thus positively and the solution negatively charged. The amount precipitated is, therefore, dependent upon the tendency of the ions to change into the metallic state. Previously only the tendency of the metals to go into the ionic condition has been mentioned ; evidently a tendency of the ions to form neutral substance, or to separate out as metal must likewise exist.

A certain potential difference must, therefore, exist at the electrode, there being some metal upon it and the corresponding ions in the solution. The magnitude of this potential difference need not be, and almost never is, the same as found when the massive metal is in contact with the solution, for the metal deposited upon the electrode does not reach the concentration of the massive metal. This conclusion seems strange at

first, for it is customary to consider the concentration
of a metal as unalterable.   This is only the case above
a definite limit.   If the metal is not present in a
molecular layer, it does not act as the massive metal.
This has been shown by Oberbeck.[1]   When the metal
of a salt solution was precipitated upon a platinum
plate the latter exhibited in the corresponding metal
solutions the potential difference characteristic of the
massive metal as soon as a certain amount had been
deposited.   Below this point the electrode exhibited
smaller potential differences corresponding to the lower
concentration.   This fact need not be surprising when
it is recalled that gases and dissolving substances have
solution pressures dependent upon their concentration.

If the source of an electromotive force be connected
with the electrode, the electrostatic attraction is counter-
acted and more ions can separate as metal.   The concentration of the metal upon the electrode is thereby
increased, and consequently also its solution pressure
(P), which tends to prevent a further deposition
of the metal, and soon entirely prevents it.   To
deposit more metal it is necessary to insert a still
greater potential difference.   This continues until the
maximum concentration of the metal is reached—that
is, until the electrode acts as the massive metal.   A
continual deposition may then take place without
further increase of the applied electromotive force, the
osmotic pressure of the ions (p) remaining unaltered.
When strong currents are used p does not remain
constant, but gradually diminishes, and consequently
the potential difference at the electrode increases.

It must be observed that the separation of the
positive ions at one electrode as neutral substance is

[1] *Wied. Ann.* xxxi. 336, 1887.

necessarily accompanied by the simultaneous deposition of the corresponding amount of negative ions at
the other.   Considerations analogous to the above
evidently apply to the negative electrode.   If, for
example, oxygen is set free, the concentration of
the gas gradually increases, and, when the solution
is saturated, has its greatest value, and consequently
its maximum solution pressure (which opposes the
further decomposition of the electrolyte).   If more
separates, it escapes into the air.   It will now be
understood why a certain electromotive force is
necessary to induce continuous decomposition in an
electrolyte : this only occurs when the concentrations
of the two substances separating at the electrodes
have reached their maximum values.   It is also
evident that the electrodes upon which metals are
deposited should exhibit the potential characteristic of
the massive metal when the decomposition point is
reached.   But it is evidently unnecessary that these
maxima of concentration for both electrodes should be
reached *simultaneously* : it may sometimes be reached
before the decomposition point of the solution can be
attained, as is the case with a silver solution, for
example.   The decomposition point of normal silver
nitrate is 0·70 volt, but the potential difference at the
electrode upon which silver is deposited is of the same
magnitude as that between massive silver and the
solution long before this decomposition value is reached.

The polarisation due to metal ions having been
considered, attention will now be directed to the
phenomena presented when gaseous or dissolved substances are separated.   These are somewhat more
complicated, and have greatly increased the difficulty
of comprehension of polarisation in general.   As a

simple case, the cell : platinised platinum in hydrogen
—an electrolyte as sulphuric acid solution—platinised
platinum in oxygen, both gases being under atmospheric pressure, will be considered. The cell at 17°
has an electromotive force of about 1·07 volt, and is
to be considered reversible. If an opposing electromotive force of 1·07 volt be connected with this cell,
a condition of equilibrium exists; when a lower
electromotive force is applied, water is produced by the
oxygen and hydrogen of the cell, and when the electromotive force of the opposing current is greater than
1·07 volt, water is decomposed. Smale[1] calculated
the temperature coefficient of this cell from the
Helmholtz formula, using the known heat of formation
of water under constant pressure (68300 cal. at 17°)
and the measured electromotive force as data :

$$\epsilon_0 \pi - Q = \epsilon_0 T \frac{d\pi}{dT}$$

$$96540 \times 1 \cdot 07 - 34150 \times 4 \cdot 24 = \epsilon_0 T \frac{d\pi}{dT}$$

$$-\frac{41500}{96540 \times 290} = \frac{d\pi}{dT}$$

$$\frac{d\pi}{dT} = -0 \cdot 00148.$$

Q is $\dfrac{68300}{2}$, since the heat effect corresponding to one
equivalent of the substance is employed. Experi-
mental determinations gave as a mean value between
0° and 68° 0·00141, which is a satisfactory agreement
with the calculated value.

---

[1] *Zeitschr. physik. Chem.* xiv. 577, 1894. On account of an error in
the original calculations, the value above given differs slightly from
that in the article referred to.

It may now be predicted that if the hydrogen and oxygen, instead of being at atmospheric pressure, be at a lower pressure, the electromotive force of the cell will be lower.   In fact, if the pressures of the gases be reduced almost to zero, the electromotive force will nearly disappear.   Under such a condition water may evidently be decomposed by currents of minimum electromotive force, it being only necessary to apply one which exceeds that of the cell itself by a very small amount, from which it is clear that the electrical energy obtainable through the formation of water from oxygen and hydrogen, or necessary for its decomposition (the two being equal and of opposite sign), may assume any magnitude from zero to a certain value dependent on the pressures of the gases or their concentrations.   The heats of formation at constant pressure, on the other hand, are independent of the pressure, and this is the most direct evidence that a simple relation cannot exist between the heat of reaction and the electrical energy obtained.   It is certainly possible in this case to calculate the amount of one of these forms of energy from a knowledge of the other when the changes of the temperature coefficient due to pressure changes are known.

That water may thus be decomposed by minimum quantities of electrical energy seems at first a contradiction of the law of the conservation of energy.   This is, however, in no wise the case.   The law referred to declares that by the reversible changes of a system from one condition to another, the same amount of work must always be done, and this condition exists in the present case.   The decomposition of water into hydrogen and oxygen at atmospheric pressure may be accomplished, on the one hand, by the application of electrical energy alone. A gas cell such as described, the gases being under atmospheric pressure, may be used, an opposing electromotive force just exceeding that of the cell being connected with it. Electrical energy alone then causes the decomposition of the water into hydrogen and oxygen at atmospheric pressure. This same result may, however, be brought about in another way. For instance, a hydrogen-oxygen cell in which the pressure of the gases is one-tenth atmosphere may be employed. The electromotive force of this cell being lower than the previous one, less electrical energy is required to produce the hydrogen and oxygen at the reduced pressure. But the work which corresponds to the difference between the two quantities of electrical energy employed must exactly suffice to compress the gases produced at one-tenth atmosphere to the pressure of one atmosphere, and thus the total work in the two cases, although done in different ways, has remained the same.

When platinised electrodes are used, the formation and decomposition of the water are reversible. At atmospheric pressure water may be decomposed by an electromotive force of 1·07 volt. If the electrodes are not platinised, the electrolysis does not take place until the electromotive force is 1·70 volt. This is that maximum for decomposition found for the acids and bases, hydrogen and oxygen being the products. It was long considered surprising that the decomposition point in the latter case was so high, notwithstanding the fact that only the partial pressure of the atmosphere is exerted upon each of the gases. Furthermore, the fact that the decomposition point was dependent upon the nature of the indifferent electrode appeared curious.

These results can now be understood. In the first place, when electrodes such as ordinary platinum or gold[1] are employed, the process is no longer a reversible one. These electrodes have too feeble absorbing power to remove the gases as rapidly as formed. With the platinised electrodes there is equilibrium between the gas dissolved in the solution, that dissolved in or taken up by the electrode, and the volume of gas surrounding the electrode. If the applied electromotive force be great enough to overcome that of the gas cell, gas separates at the electrodes, and thereby its concentration in the solution as well as in the electrode is increased. The former condition of equilibrium is soon reproduced, for the electrode yields its excess of gas to the space about it, which is considered so great that no change in the concentration is produced, and in this manner also prevents supersaturation of the liquid. The gas formed by continued decomposition of the electrolyte is thus added to the gas volume at constant concentration, and the generation can therefore always result from the same electromotive force.

The conditions are entirely different where the electrodes are gold or unplatinised platinum. These have practically no absorbent action on the gases, and

---

[1] If carbon be used as electrode, the kind plays an important part. Carbon is capable of taking up gases to a considerable extent, and this property increases its value as positive electrode of a galvanic element. In the Leclanché element, for example, hydrogen is evolved at the carbon pole, and this causes it to pass quickly from the liquid to the air, thus reducing the polarisation at this electrode. For long-continued activity of the cell the carbon is often incapable of removing the hydrogen, and polarisation is the result. If the action of the cell be stopped for a time, the hydrogen dissolved in the liquid has an opportunity to escape, and the element, becoming thus depolarised, exhibits its original electromotive force. It recovers.

there is thus no medium to bring about equilibrium between the solutions of the gases as formed in the cell and the gases in the space about the electrodes. Proceeding on this assumption, the result of a gradually increasing electromotive force opposing such a gas cell would be exactly as observed. Beginning with a low electromotive force, a scarcely perceptible decomposition of water would take place, the concentrations of the hydrogen and oxygen in the water being at first inconsiderable. At each subsequent increase of the applied electromotive force so much water at the most may be decomposed that the concentration of the gases in solution at the electrodes is made exactly that which would produce an equivalent electromotive force with platinised electrodes. A higher concentration of the gases can evidently not be produced, otherwise perpetual motion would be possible. This explains the temporary current observed in the galvanometer. Diffusion alone causes disturbances, the gases being thereby very slowly removed from the electrodes and the concentration reduced so that further decomposition takes place. The galvanometer corroborates this, since, after the first deflection, the needle does not return quite to its former position, and thus a slight current is indicated. Gradually increasing the electromotive force, the concentration of the separated gases continually increases, until finally a point is reached where gas bubbles are formed. That such an evolution of gas only occurs when the electromotive force is relatively high is explicable on the assumption that the application of considerable work is necessary for the production of the bubbles. When this point has been reached, water may be decomposed without further increase in the concentration of the solutions of the gases at the electrodes. The gases are continually evolved as bubbles, and the so-called decomposition point is observed, that is, that point above which water may be continually decomposed without the aid of diffusion. The less the diffusion of separated substance from the immediate neighbourhood of the electrode, the more marked is the decomposition point, and indeed often (in the case of metals) the galvanometer exhibits a clearly defined sudden rise in the strength of the current.

It has been seen that the decomposition point is reached when the separated gases are first evolved. This evolution takes place through the formation of bubbles at the electrodes. The process may be likened to the boiling of a liquid, and just as the ebullition does not occur at a perfectly definite temperature, but may be retarded in different ways, so also the evolution of gases as bubbles in the electrolysis is to be considered within certain limits as accidental. Some electrodes, through their physical properties, favour this evolution more than others, and thus the decomposition point is dependent upon the nature of the electrode.

Primary Decomposition of Water.—The electromotive force of the hydrogen-oxygen gas cell is dependent upon the concentrations of the gases, but nearly independent of the nature of the electrolyte. This may almost equally well be acid or base. The electromotive force is the sum of the potential differences produced at the hydrogen and oxygen electrodes. That of the former is dependent upon the concentration of the hydrogen ions, that of the latter upon the concentration of the hydroxyl ions. According to the law of mass action, the product of the concentrations of the hydrogen and hydroxyl ions is always constant without regard to other substances present; therefore, although the values of the single potential differences may vary considerably on changing the homogeneous solvent, their sum always remains the same (p. 230).

Leaving out of account metal salt solutions reducible by hydrogen, and chlorides, bromides, iodides, etc., reducible by oxygen, the ions of water alone take part in the decomposition, instead of those of the dissolved electrolyte, so that with the limitations given the law may be expressed : *In electrolysis a primary decomposition of the water takes place.* The actual electrical conductivity is brought about by all the ions in the solution, but at the electrode that action takes place which proceeds most easily, and this is the separation of the hydrogen and hydroxyl ions. When, for example, a solution of potassium sulphate is being electrolysed, and the current is not too strong, there is no reason for assuming the separation of potassium and the SO_4 radical at the electrodes, and the subsequent or secondary action of these upon the water. This assumption, though usually made, seems to the author to introduce an unnecessary complication. What is actually observed is the separation of hydrogen and oxygen ; furthermore, it has been seen that the formation and decomposition of water is a reversible process, or that in the decomposition there is no unnecessary loss of work. With the assumption of secondary decomposition such a loss should occur.

Lack of hydrogen and hydroxyl ions can never occur, since ions must be immediately generated by the undissociated water, the product of the two ion concentrations always having a definite value. After

these remarks the results obtained for the polarisation
when unplatinised electrodes are used may be understood.   The substance in presence of which the water
is decomposed will be first considered.

Acids and bases must have the same decomposition
point, because the product of the concentrations of the
hydrogen and hydroxyl ions in the solution, and consequently the sum of the single potential differences,
remains constant.   In the electrolysis of salts this
point must be higher, because at that electrode where
hydrogen separates, a base is produced.   The accumulation of the OH ions causes the concentration of the
H ions to be reduced; therefore the potential difference is increased.   Similar considerations apply to the
oxygen electrode, acid being formed, and the concentration of the hydroxyl ions thereby reduced.

The weaker the tendency to dissociate characterising
the acid or base, the less will be the rise of the
decomposition point, as is actually observed.   Since
that ion leaves the solution which requires the lowest
electromotive force for its separation, other ions than
hydrogen and hydroxyl come into account only when
the electromotive force requisite for their continued
separation is less than for these two ions.   This explains the fact that the decomposition points of the
halogen acids, which do not yield oxygen, are lower
than for acids through whose electrolysis oxygen is
evolved.   Furthermore, although the decomposition
point of those acids and bases yielding hydrogen and
oxygen is not dependent on the concentration, because
the product of the H and OH ions is constant, with
halogen acids it rises as the concentration diminishes,
owing to simultaneous diminution in the number of
hydrogen and halogen ions.   In consequence, the

number of the hydroxyl ions is continually increased, and with increasing dilution the point is finally reached where oxygen is more easily evolved than halogen. At this point the solution exhibits the decomposition point of water, as is illustrated by hydrochloric acid (p. 249).

The Significance of the Electromotive Force for Electrolytic Separations.—As already shown, different decomposition points characterise the various metals, and from this fact it ought to be possible to quantitatively precipitate metals one after another from their mixed solutions by a gradual increase in the electromotive force of the decomposing current. That this may be done has been shown by Freudenberg.[1]

If in a solution containing salts of copper and cadmium a current be employed whose electromotive force is insufficient for the continual deposition of the cadmium, but capable of precipitating the copper, this metal alone is completely precipitated. When all the

---

[1] *Zeitschr. physik. Chem.* xii. 97, 1893. About ten years ago M. Kiliani called attention to the possibility of electrolytic separations by a gradation of the electromotive force, and carried out the separation of silver and copper. He came upon the idea in considering the heat effects characterising individual metals, and calculated from them the electrical energy necessary for their precipitation. This method of calculation has been shown to be inapplicable, for which reason, and perhaps more especially because of the general uncertainty regarding polarisation conditions introduced, his work did not receive much attention. That when the electromotive force is above a certain value a metal may be continually precipitated from its solution, while below this point only an analytically negligible or absolutely unweighable amount precipitates, was not at that time clear. The opinion was then much more commonly held that even with low electromotive forces not inconsiderable quantities of the metal were precipitated, according to which view the separation of two metals by a proper regulation of the electromotive force appears as an accident rather than a necessary result of recognised relations.

copper is precipitated the current ceases, it being thus
unnecessary to pay attention to the electrolysis. The
electromotive force necessary for the precipitation of
the copper increases with the dilution of the solution,
according to the formula,

$$\pi = \frac{RT}{n_e \epsilon_0} \ln \frac{P}{p} \; ;$$

but since an increase in dilution from $\frac{1}{10}$ to $\frac{1}{1000000}$
normal (the limit of analytical determinations) causes
an increase of less than 0·3 volt for a monovalent and
half as much for a divalent metal, the separation may
usually be made complete.

After the precipitation of the copper the electromotive force may be increased and the cadmium precipitated.   In this way a number of separations have
become possible which had not succeeded when attention was given to changing the current-strength instead
of the electromotive force.   In the future this must be
kept in mind in all processes of electrolysis.

Besides the neutral or acid solutions, those of the
double compounds of the metal salt with ammonium
oxalate or potassium cyanide are especially adapted to
such separations.   In the latter many metals can be
separated from one another which cannot in acid solution.
Thus in acid solution platinum cannot be separated
from gold, mercury, and silver, *i.e.* from the metals
with slightly different solution pressures, but is easily
separated in potassium cyanide solution.   This depends
upon the formation of the complex salt $2K, Pt(Cy)_6''$,
whose negative ions are dissociated to an extremely
slight extent into Pt^{IV} and $6Cy$. As a result of
the infinitely low concentration of the ions, the platinum cannot be precipitated by the electromotive force

sufficient to precipitate the other metals whose ions are more numerous (see also p. 175).

Previously, in the quantitative separation of the metals, only the current-strength was altered. In a mixture of zinc, copper, and silver salts in acid solution the silver must separate first, since that process occurs requiring the least expenditure of work, which is also the case even though the electromotive force be very high, provided that sufficient silver ions are present at the electrode. The current must be stopped at the proper moment, otherwise the second most easily · separated metal will be precipitated. After silver and copper, hydrogen follows. To precipitate zinc simultaneously with the latter from an acid solution the current-strength must be made so great that the hydrogen ions present are insufficient to convey all the electricity from solution to electrode, and zinc ions must take part in the process. It is evidently more rational to regulate the electromotive force instead of the current-strength, and thus do away with the energy loss involved. Until within the last few years most electrolytic separations were carried out empirically without knowledge of these theoretical principles.

Synthesis of Organic Substances.—A word may be said in closing concerning the electrolysis of organic compounds, especially of the acids. A well-known example of such an electrolysis is seen in the decomposition of acetic acid into hydrogen, on the one hand, and ethane and carbon dioxide, on the other.

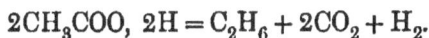

$$2CH_3COO, 2H = C_2H_6 + 2CO_2 + H_2.$$

The method is now much used for the production of certain compounds. Crum-Brown and Walker[1] ob-

[1] *Lieb. Ann.* 261, 107, 1891 ; 274, 41, 1893.

tained as principal product of the electrolysis of the ethyl potassium salt of normal dibasic acids the diethyl ester of the normal acid of the same homologous series :

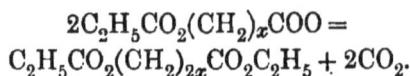

$$2C_2H_5CO_2(CH_2)_xCOO =$$
$$C_2H_5CO_2(CH_2)_{2x}CO_2C_2H_5 + 2CO_2.$$

These syntheses usually take place in concentrated solutions only, and at high current-strength. In dilute solutions with not too great current-strength only hydrogen and oxygen are evolved. This is explicable from the following consideration : At the anode (to which the anions migrate) there are OH and acid ions. As the decomposition point for the OH ions is the lower, oxygen is evolved, new OH ions are formed from the water, but this process is not infinitely rapid ; therefore, if the current be too great, insufficient OH ions are produced and the acid ions partially take their place. Increasing the concentration of the solution has the same effect as increasing the current-strength. The osmotic pressure of the acid ions being increased with increasing concentration, their change into the neutral condition is facilitated. The potassium salt is chosen instead of the free acid because of its greater conductivity.

Conceptions of Water Decomposition.—It may be here repeated that the assumption often made, in accordance with which those ions primarily separated at the electrodes are brought there in the conductivity, and that these act secondarily upon the water or other material present, does not appear to the author to accord with the facts. That the conductivity of the current through the solution and the decomposition at the electrodes do not at all stand in that close relation usually accredited to them is shown by simple consideration of the fact that in the electrolysis of any electrolyte at either electrode more ions leave the solution than migrate to the electrode through it (p. 68). Thus in every case a part of the ions originally at the electrode must be precipitated without having migrated through the solution.

The following conception is much to be preferred : Conductivity and separation at the electrode are not closely connected phenomena. All ions in the solution share in the electrical conductivity, while at the electrode those ions leave the solution which demand for their separation the least consumption of work. Therefore it happens, for example, that the ions of water which scarcely take a measurable part in the actual conductivity play the most important *rôle* in the separation at the electrodes. The assumption of secondary reactions is usually entirely unnecessary.

The following example well illustrates the simplicity of the new conception as compared with the old. Suppose a fairly concentrated aqueous solution of salts of potassium, cadmium, copper, and silver to be electrolysed between platinum electrodes through application of a not too strong current. The ions K, Cd'', H, Cu'', and Ag simultaneously come to the negative electrode. The experimental result is that only metallic silver is deposited at first. After some time, the number of silver ions at the electrode being no longer sufficient for the current density, copper also precipitates, then hydrogen, and finally cadmium. Is not the simplest conceivable expression of the results of the experiment contained in the following sentence ? Those ions which give up their electrical charges most easily are first precipitated through primary action, each other metal remaining until those preceding it in the series are

removed.   The process as thus explained becomes
simple and clear.

What of the other explanation ?   This necessitates
the simultaneous precipitation of the cadmium, copper,
potassium, and silver.   The potassium can now act
upon the water producing hydrogen, precipitate copper
from the copper salts, cadmium from the cadmium
salts, and silver from the silver salts.   (It cannot be
assumed that a silver particle is always in the immediate neighbourhood of the potassium, but the latter
would precipitate whatever metal presented itself.)   The
deposited cadmium can precipitate hydrogen from the
water, copper from the copper salts, and silver from the
silver salts ; the hydrogen can precipitate copper from
the copper and silver from the silver salts, and, finally,
the copper must precipitate the silver from the silver
salt.   This conception of the process cannot be called
simple, and why the assumption of all these secondary
reactions which no one has observed, and which are in
no sense necessary !

CHAPTER VIII

THE chemical processes taking place in galvanic elements may now be briefly considered. It will be assumed that the remarks in the chapters on electromotive force and polarisation, especially the influence of dilution (p. 230), are understood.

Constant Elements.—Besides the Daniell, the Helmholtz, Clark, and Weston elements are used as so-called normal elements (p. 128). These are :—

Helmholtz: Zinc — zinc ~~sulphate~~, mercurous chloride, mercury.

Clark : Zinc — zinc sulphate, mercurous sulphate, mercury.

Weston : Cadmium — cadmium sulphate, mercurous sulphate, mercury.

They have the advantage over the Daniell element of remaining unaltered for an indefinite time, and may be transported without incurring changes affecting the electromotive force, while the Daniell is preferably made shortly before use, because of the disturbing effects due to the diffusion of the solutions into each other. In these normal elements amalgams containing about 10 per cent of the metals, zinc or cadmium, may advantageously be used instead of the pure metal.

An H tube, into which two platinum wires are fused below (Fig. 33), is advantageously used. The amalgam is placed in one limb and mercury in the other. Some solid mercurous sulphate or chloride is placed upon the mercury, and the tube, being filled with the zinc or cadmium sulphate solution, is closed with corks covered with paraffin.

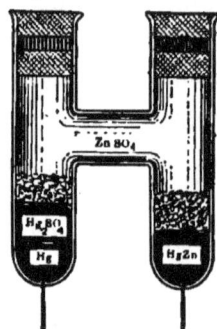

FIG. 33.

The chemical reaction taking place when these elements are active consists in the passage of the positive zinc (or cadmium) ions into the solution and the deposition of ions as metal from the solid mercurous salt at the other electrode. These elements differ from the Daniell in that they can only produce very feeble currents. On account of the difficult solubility of the mercurous salts used, the quantity of mercurous ions is very small, and the replacement of these ions by the dissolving of more mercurous salt takes place but slowly. On this account the electromotive force of the element rapidly diminishes when too much is required of it. The simultaneous supersaturation of the positive ions at the negative electrode also tends to reduce the electromotive force. If the element be allowed to stand, the original condition is again attained, or the element recovers.

In the Clark or the Weston element, where the zinc or cadmium solution as well as that of the mercurous salt is saturated, the conditions before and after normal activity differ in that the quantity of amalgam is slightly smaller, while the solid zinc or cadmium sulphate is slightly greater, the mercurous sulphate being diminished in amount and the quantity of pure mercury somewhat augmented. Practically, then, the zinc or cadmium ions entering the solution receive an equivalent amount of SO_4^{II} ions through the solution of the solid mercury salt, and solid zinc or cadmium salt is found. The zinc or cadmium and the mercury ions remain unaltered in concentration so long as amalgam and solid mercurous salt are present; therefore the cells are strictly constant. The same applies to the Daniell element when saturated solutions in contact with the solid salts are used. In the Helmholtz cell, on the contrary, the zinc chloride solution increases in concentration when used, and thereby a change, though practically an inconsiderable one, takes place in its electromotive force.

The changes of electromotive force with the temperature must finally be considered. The composition of the solutions remaining unaltered, the temperature coefficient of an element is practically the sum of the coefficients of the potential differences at the two electrodes. In these elements, where saturated solutions containing also an excess of the solid salt are employed, the change of the solubility of the salt with the temperature plays an important part. The relatively high temperature coefficient of the Clark element is principally due to the fact that with rising temperature the solubility of the zinc sulphate greatly increases, that is, the concentration of the zinc ions becomes greater. The solubility of cadmium sulphate is only slightly influenced by the temperature, and the temperature coefficient of the Weston element is nearly zero.

Inconstant Elements.—In the elements just described the constancy of the electromotive force only was of importance, the height of this and the cheapness of the materials being of very secondary importance.   With the elements necessary for the ordinary electrical work in the laboratory these relations are nearly reversed.   High electromotive force combined with economical processes is here more important than great constancy.   The number of different kinds of galvanic elements in use is very great; three of them may be profitably studied here.

Formerly the cell containing zinc, ammonium chloride solution, bleaching powder solution, carbon, was much used.   The ammonium or sodium chloride solution is separated from that of the bleaching powder by a porous earthenware cell.

Through the action of this element, zinc ions come into existence at the negative electrode and chlorine at the positive.   The bleaching powder supplies the chlorine ions and the carbon acts as conductor.   The electromotive force is at first very considerable, because the concentration of zinc ions in the chloride solution is extremely small, while the solution pressure of the chlorine in the bleaching powder is great.   When the element is being used the concentration of the zinc, as well as of the chlorine ions, increases, and both changes reduce the electromotive force.   The solution pressure of the chlorine remains constant so long as any of the solid bleaching powder remains.

If the porous cell and the bleaching powder of this element be removed, and the carbon be replaced by a mixture of carbon and manganese dioxide, the much-used Leclanché element results.   Distinction must here be made between the action of the dioxide and of the carbon ; for the present the former will be left out of account.   If the element be closed, zinc goes into solution, and hydrogen ions, being more easily deposited than the NH_4 ions, which participate in the conductivity, separate at the carbon electrode.   Carbon has a considerable solvent action upon gases, and rapidly conducts the separating hydrogen into the air, thus preventing the accumulation of hydrogen dissolved in the water.   An accumulation of the gas at the electrode, by rendering difficult the separation of more hydrogen ions, would cause a reduction of the electromotive force.   This, in fact, occurs when the element is allowed to work too rapidly, the ability of the carbon to remove the hydrogen being overtaxed.   If the element be allowed to stand inactive a short time it recovers.

The manganese dioxide aids the carbon, as is evident from the following considerations.   Since every substance has a certain solubility, such must be ascribed to the dioxide, and Mn^{IV}, together with the corresponding OH ions, may be considered present in the solution. The quadrivalent manganese ions tend to give positive electricity to the electrode and become bivalent. Therefore, while zinc ions are formed at the negative, the corresponding amount of manganese ions change their valency from four to two, and manganous chloride is formed at the positive electrode, the Mn^{IV} ions being replaced from the solid MnO_2.   Which of the processes described predominates in the element depends upon the composition of the mixture of carbon and dioxide.   By long-continued use the electromotive force diminishes, principally because of the accumulation of zinc ions.   This may be remedied by renewing the ammonium chloride solution.

Another frequently used element having a high electromotive force is the so-called bichromate element, containing zinc — chromic acid (or sodium bichromate with sulphuric acid) — carbon. Zinc ions are formed at the negative pole as usual, but the reaction taking place at the other electrode is more complicated. It may be assumed that the ions Cr_2O_7'' are present, the chromium being sexivalent. As in the case of the compound H_2PtCl_6 it was assumed that the negative ions $PtCl_6''$ were slightly dissociated into the quadrivalent platinum and univalent chlorine ions (p. 175), so also here the Cr_2O_7'' ions may yield a minimum quantity of sexivalent chromium and the corresponding quantity of univalent OH ions. The chromium ions of high valency tend to change into ions of lower valency, probably trivalent, and the high electromotive force, with the exception of the potential difference at the zinc electrode, depends upon this change. The number of zinc as well as trivalent chromium ions increases with the time, while the concentration of the sexivalent chromium ions decreases, and each of the three changes should cause a reduction of the electromotive force. The electromotive force of the active bichromate element does actually diminish rapidly.

ACCUMULATORS

Accumulators are arrangements in which electrical may be stored as chemical energy, and whence it may again be obtained at wish in the form of electrical energy. Any reversible cell may be used as an accumulator. If a current be sent through a used Daniell element in the direction from copper to zinc, copper is dissolved and zinc precipitated—in other words, electrical energy is stored up in the form of chemical. In practice lead accumulators are used almost exclu-

sively.[1]   The electrodes consist of lead plates coated
with a specially prepared layer of lead oxide or sulphate,
and the electrolyte is 20 per cent sulphuric acid.
When a current is sent through this arrangement, lead
superoxide (or a corresponding hydrate) is formed on
that electrode at which the positive electricity enters
the acid, while at the other electrode metallic lead in
spongy form is produced.   The accumulator is thus
charged after the conduction of sufficient electricity
through it.   In the discharge both the superoxide and
the metallic lead return to sulphate.   The chemical
process on charging is then essentially the change of
lead sulphate to lead at one electrode, and to superoxide
at the other, while the discharge is simply the return
of these substances to lead sulphate.   The corresponding heat of reaction is given by Streintz [2] as follows :

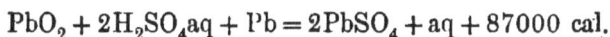

$$PbO_2 + 2H_2SO_4aq + Pb = 2PbSO_4 + aq + 87000 \ cal.$$

If the electromotive force of the accumulator be
calculated from the known heat of reaction, assuming
complete transformation into electrical energy, $1\cdot885$
volt is obtained.   This agrees very well with the
experimentally determined value for dilute sulphuric
acid of $1\cdot900$ volt.   From this agreement it also
follows that the electromotive force of the accumulator
is nearly independent of the temperature (p. 142), and
this has also been demonstrated by Streintz.   It is
therefore very probable that the reaction takes place
as represented above.

The process as yet has not been explained on the

[1] For particulars concerning the making and use of accumulators
attention is called to the work of Heim, *Die Accumulatoren*, Leipzig,
Oskar Leiner, and that of Elbs, *Die Accumulatoren*, Leipzig, Johann
Ambrosius Barth.

[2] *Wien. Akad. Ber.* 103, Jan. 1894.

basis of the ion theory ; the following is an attempt in this direction.

The accumulator being charged and ready for use, the positive electrode is coated with superoxide of lead and the negative with the spongy metal ; between the two is sulphuric acid.   It was lately pointed out that in the Leclanché element the manganese dioxide in contact with the water produces quadrivalent manganese and the corresponding OH ions.   Analogously quadrivalent Pb^{IV} ions must be formed at the positive electrode of the accumulator, and just as the quadrivalent manganese ions in the Leclanché element are changed to bivalent, so here the quadrivalent lead ions also change into bivalent.   *This process is the principal source of the electromotive force of the accumulator.*   The quadrivalent lead ions disappearing are continually supplied by the solid superoxide.   The bivalent lead ions formed, instead of remaining in the solution, combine with the SO_4'' ions to form solid lead sulphate, since this is difficultly soluble, that is, the value of the concentration product of Pb'' and SO_4'' ions is small.

At the negative pole metallic lead changes into bivalent ions, a process taking place without producing any considerable potential difference.   Here also insoluble lead sulphate is formed from the Pb'' and SO_4'' ions.

Moreover, the ion theory not only renders clear the changes of superoxide and metallic lead into sulphate, but explains the gradual diminution of the electromotive force of the accumulator in action.   The magnitude of the potential at the positive electrode depends upon the concentration of the quadrivalent and bivalent lead ions in the presence of excess of metallic lead. The concentration of the quadrivalent ions decreases with time, and that of the bivalent increases, as may be seen from the following. At the superoxide electrode there is a saturated solution of this compound—that is, the product of the concentration of Pb^{IV} and the fourth power [1] of the concentration of the OH ions is a constant. On the other hand, there must be definite relations between these concentrations and those of the sulphuric acid ions. The product of the concentration of the H and OH ions in the solution must have a constant value equal to that of water. It has been seen, in the first place, that during the discharge of the accumulator, lead sulphate is formed at the superoxide electrode, and in the second, that newly formed OH ions produced by the superoxide cannot exist as such, but must combine with the H ions of the acid to form water. There is thus a continual removal of H and SO_4'' ions taking place. The removal of the former allows of an increase in the concentration of the OH ions, and therefore causes a reduction in that of the quadrivalent lead ions. The removal of SO_4'' ions allows of an increase in the concentration of the Pb'' ions, since the solution is saturated with lead sulphate. This latter process also takes place at the negative electrode. When the supply of superoxide is exhausted, the electromotive force falls very rapidly to an exceedingly low value.

After the accumulator has been discharged there is lead sulphate on both electrodes, consequently bivalent lead ions are present. The process of charging consists simply in the change of bivalent lead ions to quadrivalent at that electrode at which the positive electricity enters the solution, and to metallic

---

[1] Because four OH ions correspond to one of the lead ions.

lead at the other electrode. The Pb^{II} ions used are replaced from the solid lead sulphate. The Pb^{IV} ions and the OH ions present, having reached that concentration in the solution determined by the dissociation constant for superoxide of lead, combine to form this oxide (or a hydrate). Thus the lead sulphate at one electrode gradually changes into superoxide, and into metallic lead at the other. The opposing electromotive force of the accumulator increases during the charging, because the processes described as taking place during discharge are reversed. The concentration of the bivalent lead ions at both electrodes diminishes with time, while that of the SO$_4$^{II} ions is continually increasing. The concentration of the Pb^{IV} ions increases with the increase of H ions formed with equivalent quantities of OH ions from the undissociated water. The OH ions continually combine with the Pb^{IV} to form superoxide, and their concentration must diminish as that of the hydrogen ions increases. The lower the concentration of the OH ions the greater is that of the Pb^{IV} ions. If no more bivalent lead ions are present, the hydrogen ions separate at one electrode and hydroxyl ions at the other. Thus the rapid generation of hydrogen and oxygen at the electrodes in charging shows that the accumulator is overcharged. In order to cause a considerable generation of hydrogen and oxygen in the accumulator, a somewhat higher electromotive force is required than is necessary to charge it, since the separating gases can accumulate to a high degree of concentration, owing to the existing conditions; otherwise the charging of the accumulator could only be brought about with a great loss of electrical energy.

SUBJECT INDEX

LIST OF AUTHORS' NAMES

THE END

Printed by R. & R. Clark, Limited, *Edinburgh*